IN RUST WE TRUST

Tess Marshall

First published in Great Britain in 2021

Editing, design, typesetting and publishing by UK Book Publishing

www.ukbookpublishing.com

ISBN: 978-1-914195-20-4

About the book

This text book is intended to be a reference text for all those working in, and with an interest in, corrosion protection by coatings. It will be particularly valuable as a reference for plant engineers and maintenance managers, as well as contractors and inspectors. There is also a detailed opening chapter on corrosion theory, and the different types of corrosion, which sets the scene for the rest of the book.

There is a detailed chapter on surface preparation, why it's needed, and the different methods employed, including a very detailed section on abrasive blasting equipment. Similarly, the chapter on paint application includes information on the many methods used, and a detailed section on airless spray equipment.

There are chapters on the different types of protective coatings used, and on the defects found in the applied coating. The relevant international standards are also referenced throughout the text.

About the author

Tess Marshall received his first coating inspection qualification in 1993. Since then he has spent 27 years in the global coatings industry, having been involved in some of the world's most demanding coating projects. He is a NACE Protective Coatings Specialist PCS2, NACE Senior Corrosion Technologist, NACE Level 3 Coatings Inspector Peer Reviewed, and is also a Professional Member of the Institute of Corrosion. He is now working as a coatings consultant within the oil and gas industry. Mr Marshall lives in Newcastle, UK with his family.

About the editor

The book was edited by Brian Goldie, who has more than 30 years' experience in protective coatings, both practically and commercially.

CONTENTS

CHAPTER ONE
Introduction to Corrosion

Corrosion is an electrochemical process occurring when a metal is exposed to the environment which changes the properties of the metal resulting in degradation. Corrosion occurs in stages as stated below:

- The ferrous ions go into solution at the anodic sites
- Electrons are released from the anodic sites and pass through the metal to the cathodic sites.

According to the National Association of Corrosion Engineers (NACE) International, corrosion is defined as "the deterioration of a material usually a metal because of a reaction with its environment".

Corrosion of a steel substrate involves:

- The combination of oxygen and moisture to form hydroxyl ions
- Ferrous and hydroxyl ions react to form ferrous hydroxide
- Ferrous hydroxyl further oxidizes to produce hydrated ferric oxide.

What is Corrosion

It is commonly thought that corrosion is the same as rust, which is not entirely true. Rust is the corrosion product formed when iron and steel corrode; however, many other materials are also attacked. The patina on copper and white rust on zinc surfaces (zinc salts) are well known examples, which are also corrosion products. Besides destroying metals, corrosion also contributes towards the breakdown of concrete and plastic materials.

In the manufacturing of metals, energy is needed to extract the metal from its ore. During the process the iron ore melts and the metal is separated from the slag products and milling takes place while the metal is in a red-hot state. During milling this red-hot steel comes into contact with oxygen and a thin layer of "mill scale" is formed on the surface.

Mill scale is a brittle oxide film measuring only microns in thickness and only protects the metal as long as it is intact. If the steel is left outside to "weather" the mill scale will eventually crack and disbond due to differences in thermal expansion leaving the substrate to rust. Chemically speaking rust has almost the same composition as ore.

The corrosion process has now come to a close as it started off with "ore" and now ended up with "rust". To extract the metal found only in limited quantities in the ore, energy is required. Due to the energy expended in obtaining the metal it is transferred from its stable level to a non-stable level.

The corrosion process is a natural process which under the right circumstances (oxygen and moisture) tries to convert the steel back to its most stable state, which is iron ore. The chemical compositions of ore and the corrosion products are very much alike. In order to prevent the corrosion process from starting we use protective coatings and linings or other barriers to stop any direct contact between the metal and the environment.

Corrosion Theory

In order for a metal to corrode the surface of the metal must be in contact with an electrically conductive liquid (electrolyte) and oxygen. Water, either fresh water or sea water, are examples of electrolytes and pollution in the air may dissolve in water making it more conductive and therefore accelerating the corrosion process. Metals have a tendency to release electrically charged atoms (ions) when in contact with water.

For some metals, this tendency is weak, for example in the case of platinum or gold as such metals are called "noble metals". Other metals release a large number of ions in contact with water for example aluminium and zinc, such metals are called ignoble metals. If we immerse two different metals in a conductive liquid (electrolyte) without any electrical connection between them, both metals corrode at a certain rate. We say they have different potentials.

However, if the metals are joined together electrically, the rate of corrosion increases on the less noble metal. How fast the corrosion takes place depends on a number of factors, including the metal types, humidity and pollution on the surface. A simple explanation of the corrosion process is provided by the activity within a dry cell battery. The dry cell battery consists of a carbon or graphite rod in a substance able to conduct electricity (an electrolyte).

This substance is enclosed by a zinc case with the carbon or graphite rod being more noble than the zinc case. If a small light bulb is inserted between the poles of the battery it will light up as the battery supplies the power. When in use the battery is being discharged, this is because the zinc case or anode corrodes. If this process is allowed to continue long enough the battery will eventually stop working and could start to leak.

The corrosion process has corroded the zinc case surrounding the electrolyte consequently the less noble of the two metals in contact has broken down and being the anode, zinc has delivered the energy to the cathode. In order for corrosion to take place it is not necessary that dissimilar metals are joined. It is a well-known fact that steel corrodes if exposed to the elements of wind and rain with the reason being is that steel is not microscopically uniform. It contains a number of small areas which are anodic or cathodic to each other.

On unpainted steel, mill scale, graphite grains, or other noble particles tend to become small cathodes, while the "steel surface" becomes the cathode. These areas form among themselves small corroding "batteries". The corrosion attacks ignoble areas on the steel (anodes). At the cathode there is no breakdown of the metal, but the solution around the cathode becomes alkaline.

It is important to note that if we can prevent the metal surface from getting into contact with oxygen and moisture, we will be able to prevent the corrosion process.

Elements of a Corrosion Cell

- **Anode:** An Anode is an electrode to which electric currents flow into a circuit. In electrochemistry the anode is where oxidation occurs and is the positive polarity contact in an electrolytic cell. The electrons flow from the anode into the cathode with corrosion occurring at the anode.
- **Cathode:** A Cathode is an electrode through which electric currents flow out of a circuit. The direction of electric current is by convention opposite to the direction of the electron flow. The electrons flow into the polarized electrical device to the anode. The cathode is protected from corrosion.
- **Electrolyte:** An electrolyte is any substance or liquid that is electrically conductive and is contact with the substrate. An electrolyte containing chlorides is more conductive than an electrolyte without any chlorides. Sea water is an excellent example of an electrolyte and is also very conductive.
- **Metallic Pathway:** The conductor through which current is carried, will be a metallic object or structure as this provides the way for electrons to flow. The metallic pathway could be a ship's hull or exposed steel work on an offshore oil rig or platform.

For example there are so many ways in which we can prevent metal surfaces from getting into contact with moisture (the electrolyte). Surfaces can be painted or applied with other types of protective coatings. But it is also practically possible to remove or reduce humidity in Closed-in constructions to such an extent that the formation of a conductive water film is not possible.

This principle is used worldwide on bridges made as steel box girder constructions. The outside surface of the steel construction can have a duplex system applied, typically a thermally sprayed zinc coating and paint but inside the steel is only primed. A dehumidifying system ensures that the relative humidity inside the structure is constantly 40%. The idea is that in such a low relative humidity the formation of an electrolyte is not possible, this prevents any internal corrosion.

Galvanic Series

A galvanic series is simply a listing of metals in the order of their potential in a specific electrolyte. Galvanic corrosion affects the anode of the couple by increasing its corrosion rate. The results of this corrosive attack may be by general attack, pitting or some other form of corrosion. Galvanic activity does not control the form of corrosion which occurs at the anode, it only increases the rate of attack. Galvanic corrosion affects the cathode of the couple in several ways.

In general the corrosion that would normally occur if the cathode were exposed to the environment by itself will decrease when it is coupled to a more active material. In some cases, however, the cathode can be adversely affected. In addition, the alkaline cathodic reaction products formed in a galvanic cell can adversely affect some materials such as aluminium.

Metals that are chemically vulnerable in both acidic and alkaline environments known as amphoteric materials can be affected at the cathode in a galvanic corrosion cell. Such attacks are called amphoteric effects. The potential difference between materials causes current flow when dissimilar materials are coupled in a suitable electrolyte. The direction of current flow depends on which metal is more active.

The more active metal becomes the anode, while the less active metal becomes the cathode. The magnitude of the driving force for galvanic attack in this couple is the difference in potential between these electrodes. To determine the specific affects between specific materials in a specific environment it is necessary to know the galvanic series for those materials in the environment of interest.

Such an empirically developed galvanic series of metals and alloys may be useful for predicting galvanic relationships of metals and alloys according to their potentials, measured in a specific electrolyte. It allows one to determine which metal or alloy in a galvanic couple is more active in the environment of interest.

The difference in measured potential between two metals or alloys in this galvanic series can give a prediction of the probable magnitude of corrosive effects. Most commonly when someone mentions the galvanic series, they are referring to a table which has been constructed from measurements in seawater. With certain exceptions, this series is broadly applicable in other natural waters and in uncontaminated atmospheres.

Galvanic Series

Power of Hydrogen

Indoors in a dry environment there is little danger of corrosion because there is no continuous electrolyte formed on the surface. In the case of outdoor exposure the corrosion attack depends on the environment. The more polluted the environment the faster the corrosion attack will be. Many of the industrial gases emitted into the atmosphere are transformed into acids or alkalis. Such solutions attack metals as well as other construction materials.

In order to indicate whether a solution is acidic or alkaline, the pH value of the solution is measured. The pH value is usually measured within a range from 0 to 14. Pure water has the pH value of 7 which is neutral. If a pH level of less than 7 is measured it is an acidic solution; the closer to 0 the more acidic the solution. Rainwater often has a pH value of around 5 because it has taken up acidic pollution.

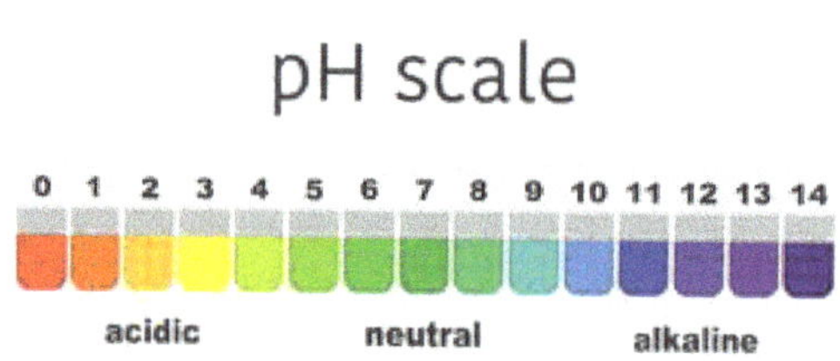

The pH scale is logarithmic, this means that for every single point above or below a pH value of 7 the solution will be 10 times more concentrated. A solution with a pH value of 5 is consequently 10 x 10 = 100 times more concentrated than a neutral solution with pH value 7. Alkaline substances have pH values from 7 to 14 with examples of such substances being sodium hydroxide, ammonia and cement.

Acidic substances such as hydrochloric acid and sulphuric acid are used in the pre-treatment of steel before hot dip galvanizing. The corrosion attack on metals is often intensified in acidic and alkaline solutions. Both aluminium and zinc sustain very severe attacks in contact with alkalis; however, if steel is immersed in an alkaline solution the steel is passivated.

A thin iron oxide film is formed on the surface which offers good protection against corrosion and the rate of corrosion is reduced considerably. Reinforcing steel bars are well protected and kept passivated by the alkaline environment in concrete. The decisive factor for the corrosion process is whether water from rain, dew or condensation due to a high relative humidity of the air is present on the steel surface.

When it rains, contaminants like salts and dirt collected on the surface will be washed away, but in the case of high relative humidity or condensation on the steel such contaminants are not washed away. Contamination in the form of dust and salts attract moisture (hygroscopic) and this prolongs the time of wetness. For this reason structures in an environment of high humidity, but little precipitation may sustain severe corrosion attacks.

Types of Corrosion

Pitting Corrosion

Pitting is a deep narrow corrosive attack which often causes rapid penetration of the substrate thickness. Pitting corrosion is characterized by corrosive attack in a localized region surrounded by corrosion free surfaces, or surfaces that are attacked to a much lesser extent. Very often a local cell is set up between the interior of the pit and the external surface.

Pitting Corrosion

The interior contains acidic, hydrolysed salts that are quite corrosive compared with the bulk solution. An anode is established within the pit and the surrounding surfaces act as cathodes. This is particularly the case for alloys that rely on a resistant passive film for protection such as stainless steels and aluminium. It also may occur on iron, steel and other metals. Pits develop at weak spots in the surface film and at sites where the surface is damaged mechanically under conditions where self-repair will not occur.

The practical importance of pitting depends on the thickness of the steel and on the penetration rate. The rate usually decreases with time, thus on thin sections pitting may be serious, while on a thick section it may be less important. In general, the rate of penetration decreases if the number (of pits) increases.

This is because adjacent pits have to share the available adjacent cathodic area which controls the corrosion current that can flow. Movement of the solution over a metal surface often reduces and even may prevent pitting that otherwise would occur if the liquid were stagnant (such as with some austenitic stainless steels in seawater).

Pit depth can be measured using a pit depth gauge. This has a needle probe. Firstly, the gauge is zeroed on a flat surface then the needle is inserted into the pit and the pit depth is read. It is important that the needle reaches the deepest part of the pit. This can be a problem with deep narrow pitting.

Crevice Corrosion

Crevice corrosion is a form of localized attack in which the site of attack is an area where free access to the surrounding environment is restricted. As crevice corrosion is caused by differences in the concentrations of materials inside and outside the crevice, crevice corrosion is also called concentration cell corrosion. This form of localized attack can occur at crevices where materials meet in such a way that the environment can enter the joint between them, but the flow of material into and out of the joint is restricted.

Crevice Corrosion

These crevices may either be metal to metal or metal to non-metal. Crevices can also be formed under deposits of debris or corrosion products. Crevice corrosion can be recognized by the location of the attack either at the entrance to a crevice or deep within the crevice. When corrosion occurs deep within the crevice it is often revealed after a failure has occurred. In oxygen concentration cell corrosion, the difference in oxygen concentration between the areas inside and outside of the crevice causes a potential difference between these areas.

Oxygen concentration primarily affects the activity of the cathodic reactions involved in corrosion. Where the cathodic activity is high, the area will act as a cathode with respect to areas where the cathodic activity is low. In many natural environments the most common cathodic reactions are the reductions of oxygen and water.

Filiform Corrosion

Filiform corrosion is a special form of oxygen cell corrosion occurring beneath organic or metallic coatings on materials. The attack results in a fine network of "threads" of corrosion products developed beneath the coating material. Such attack can develop beneath coatings in an environment of humidity higher than 60% relative humidity. Such items may include coated cans, office furniture, cameras, aircraft structures and a host of other household and common products.

Filiform Corrosion

The mechanism of filiform corrosion is similar to that of crevice corrosion in that it is driven by the potential difference between the advancing head of the attack and the area behind the head of the advancing attack. In filiform corrosion the head of the advancing filament becomes anodic with a low pH and a lack of oxygen, as compared with the cathodic area immediately behind the head, where oxygen is available through the semi-permeable coating film.

Corrosion proceeds as the cathode follows behind the anodic head. Water and oxygen in the cathodic area convert the anodic products to the usual oxides of the metal. The cause of filiform corrosion appears to be associated with mild surface contamination of solid particles deposited from the atmosphere or residue on the metal surface after processing. Such surfaces exposed to humid atmospheres will often suffer filiform corrosion.

Filiform corrosion does not normally result in significant corrosion of the metal surface. It can affect the adhesion of a paint film, resulting in the eventual attack of the metal. Filiform corrosion can occur under semi-permeable coatings on steel, zinc, aluminium and magnesium alloys exposed to humidity greater than 60% relative humidity. This type of corrosion, particularly on painted surfaces, can be prevented by proper cleaning and preparation of the metallic surface, and then applying the coating only to a thoroughly cleaned and dried surface.

Galvanic Corrosion

The mechanism of galvanic corrosion is the classical electrochemical cell. As discussed previously the electrochemical cell requires four factors to be present:

- The anode is where the metal is lost and electrons are produced.
- The cathode is the place where electrons produced at the anode are consumed.
- The metallic path conducts electrons from the anodic sites to the cathodic sites.
- The electrolyte provides reactants from the cathodic. reactions and allows the flow of ions.

For galvanic corrosion to occur all the above components must be present and active. In galvanic corrosion, electrons flow through a metallic path from sites where anodic reactions are occurring to sites where they allow cathodic reactions to occur. Electrical current flows through the electrolyte to balance the flow of electrons in the metallic path.

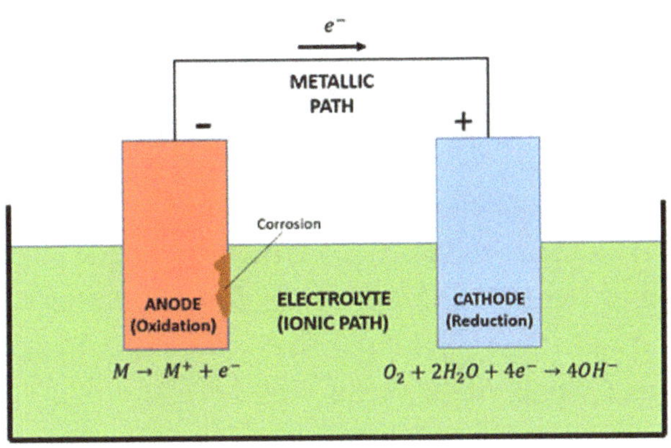

Example of a Corrosion Cell

Galvanic corrosion is driven by the potential differences between metals or electrically conductive non-metals when exposed to an electrolyte. Non-metals include such electrically conductive materials as graphite. In some cases the cathodic member of a galvanic couple can be a deposit formed by the deposition of metal ions from solution onto the anodic metal surface.

The Importance of Removing Mill Scale

As the steel is hot rolled and it exits the rolling station at temperatures of 1,250°C the steel reacts with oxygen as it starts to cool down and produces mill scale, an oxide layer which turns black over time due to a reaction with the atmosphere.

- Mill scale is magnetic and contains three layers of oxide.
- The layer closest to the steel substrate is ferrous oxide.
- The second layer is a chemical formula Fe_3O_4.
- The thin outer layer of oxide is Fe_2O_3.

Mill Scale

The thickness of the mill scale varies depending on many factors; however, it has been known to be 250µm in thickness and is quite brittle. This oxide layer is also unpredictable being tightly adherent or lightly adherent to the substrate hence greater or lesser blast cleaning times are required. Blast cleaning is the only option to remove such a tenacious oxide layer as methods such as mechanical wire brushing will only polish or burnish the mill scale.

- Higher blasting pressures may be needed to remove mill scale (above 100 psi – 7 Bar).
- Sharp and angular abrasives such as chilled iron grit are preferred.
- Greater blasting times may be required to remove such a tenacious oxide.

Mill scale is highly cathodic to bare steel so if the mill scale is not removed the steel will sacrifice itself to save the mill scale. If the steel work is stored on a quayside or a corrosive environment, pitting corrosion of the underlying surface may be found once the mill scale breaks away; this can be very dangerous as it may penetrate the steel thickness. If a steel surface is to be exposed to corrosive conditions the mill scale must be removed from the substrate to ensure the integrity of the coating system.

Applying Coatings over Mill Scale

Applying a coating over mill scale is bad painting practice and a recipe for coating failure. Mill scale is smooth and has no surface profile, hence adhesion is compromised when painted over. Eventually the mill scale will lose adhesion from the surface and flake away taking the coating with it. Mill scale has a lower thermal expansion than steel and cracks on cooling, which also allows moisture to penetrate the oxide layer so if it is painted over it will only be a matter of time before coating failure occurs.

- Mill scale is smooth and provides no profile or key for coating adhesion.
- Could be lightly adherent or tightly adherent to the substrate.
- Mill scale is cathodic to bare steel and causes pitting.
- Power wire brushing will not remove mill scale.

Pre-Surface Preparation

Surface imperfections which may be encountered during the pre-blast inspection are laminations or shelling, weld spatter, slag, undercutting, sharp projections, all of which can cause premature coating failure if not removed. The standard below describes surface defects which may be encountered before the application of a protective coating system.

The ISO 8501-3 Preparation of Steel Substrates before the application of Paints and Related products - Visual assessment of Surface Cleanliness - Preparation Grades of Welds, Cut edges and Other Areas with Surface Imperfections.

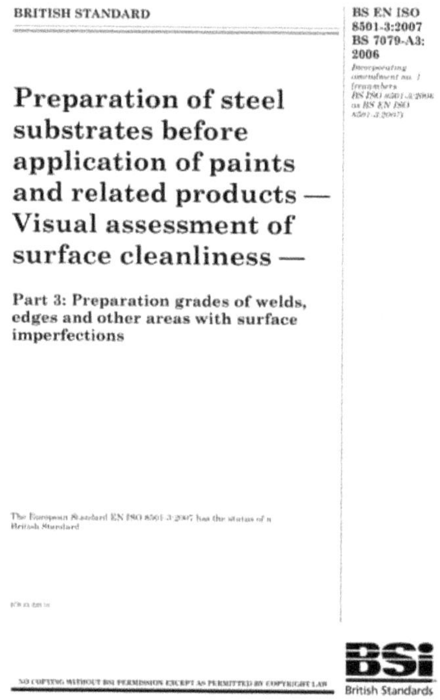
BRITISH STANDARD

BS EN ISO 8501-3:2007
BS 7079-A3: 2006

Preparation of steel substrates before application of paints and related products — Visual assessment of surface cleanliness —

Part 3: Preparation grades of welds, edges and other areas with surface imperfections

BSi British Standards

ISO 8501-3:2007

The standard defines the preparation grades of welds, cut edges and general imperfections that are not rectified by blast cleaning. Most surface imperfections are removed by hand or power tools such as hammer and chisel or grinder. If surface laminations are removed by grinding after the blast cleaning process then the surface will have to be prepared again to restore the surface profile.

Weld Spatter

Weld spatter is essentially droplets of molten metal that are generated at or near the welding arc. Spatter is easy to identify as they are small, round ball-like substances when they solidify, and are found next to the welded area. Spatter is generally regarded as a nuisance and must be fully removed prior to the application of protective coatings as premature coating failure is inevitable due to the coating flowing away and becoming thin.

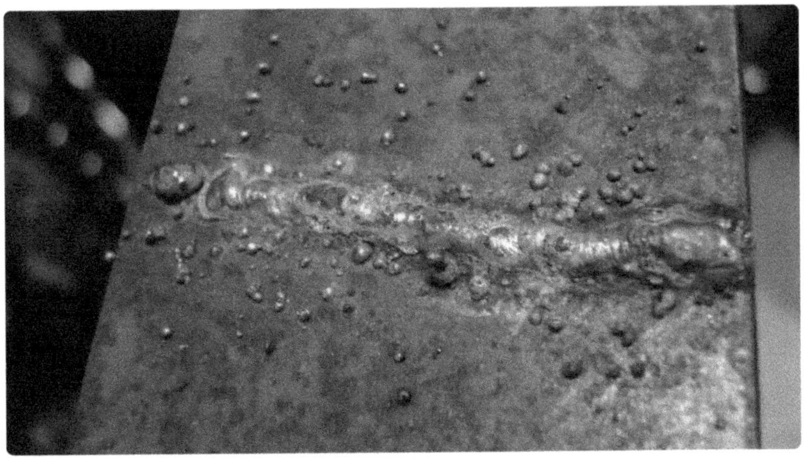

Weld Spatter

Small crevices also develop in and around the base of the spatter, which creates an area where coatings do not tend to penetrate well. These rough areas of weld spatter should therefore be removed before the surface is prepared for coating by means of mechanical tools such as grinding or chipping hammer as weld slag and spatter are firmly adhered to the substrate and blast cleaning alone will not remove them.

If weld spatter is identified after blast cleaning the area, it must be removed and the area re-blasted to restore the surface profile. The coating and welding inspector will both be checking for steel surface defects such as weld spatter.

Laminations

Laminations are formed during the mill rolling process and are sometimes referred to as surface shelling or hackles. During this process carbon steel may pass through the rollers many times to achieve the desired thickness. Each pass results in an increase in the area of the plate and decreases in thickness.

Any discontinuities present in the metal such as porosity or inclusions trapped in the piece of steel are elongated in the direction of the rolling and results in laminate formation within the metal sheet. Laminations may sometimes only be visible after blast cleaning and once identified they can be removed by hand scraper or grinding.

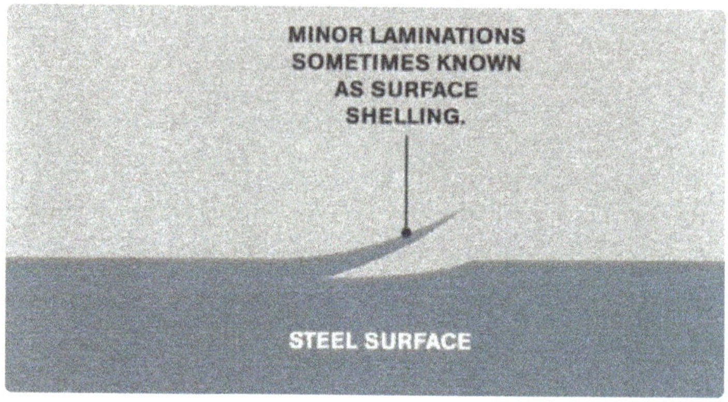

Lamination

If they are removed by the grinding/scraping process the area will have to be blast cleaned again to restore the surface profile. If laminations are left on the surface and painted over, failure is inevitable as the coating will shrink and pull back over the lamination resulting in spot rusting together with premature coating failure.

Sharp edges

Sharp edges of steel on over lapping plate or edges left by flame cutting can present serious coating problems as coatings will fail first in areas with sharp edges without exception. These should therefore be removed prior to surface preparation. A 2mm radius is generally recommended for good painting practice. The edge can be ground or radiused with a mechanical grinder; however, if the edge is ground after blast cleaning the surface profile will be removed.

Flame Cut Edge

Good painting practice dictates that the application of coatings by brush (stripe coat) directly to the edge before each full spray coat is carried out so the surface is suitably "wetted" and film thickness is increased. Not only does the surface tension pull the coating away, it also creates a thin film on the edge. In many cases the best protection is achieved by increasing the coating thickness on the edge; however, the radius requirements will be found in the coating specification.

Undercutting

Undercutting is a groove or crater that occurs near the toe of the weld. When this weld flaw occurs, the weld metal fails to fill in the grooved area, resulting in a weak weld that is prone to cracking along the toes. One reason for this type of defect is excessive current, causing the edges of the joint to melt and drain into the weld leaving a depression along the length of the weld.

Undercutting

Another reason is if a poor welding technique is used it does not deposit enough filler metal along the edges of the weld. Incorrect filler metal is also another reason as it will create greater temperature gradients between the centre of the weld and the edges. Other causes include too small an electrode angle, a dampened electrode, excessive arc length, and slow speed.

Skip Welds

Skip welding or intermittent welding is a process of welding two or three inches, skipping several inches and then applying additional welds of a similar length, until the designated area is covered. The use of skip welds can reduce distortion, speed up production, reduce cost and reduce weight. Skip welding, however, creates many problems since corrosion can occur in the area between the weld and segment not welded.

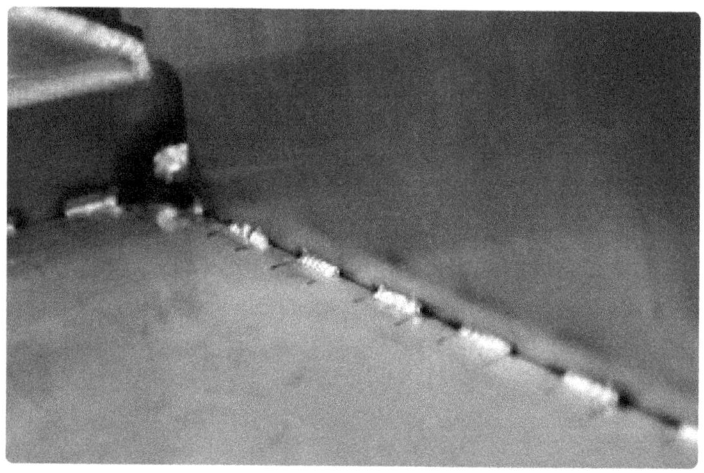

Skip Welding

This creates an issue for the application of coatings and it also allows the accumulation of contaminants such as chlorides and nitrates which can be a focal point for corrosion. If continuous welding cannot be performed and skip welding is chosen, a good alternative would be to use a compatible mastic or caulking compound to fill in the areas between the skip welds, reducing the build-up of corrosion prompting products.

Blow Holes

Blow holes or porosity refers to cavity-type discontinuities or pores formed by gas entrapment during the solidification of molten weld metal. Porosity reduces the strength of a weld. In arc welds it is caused by dissolved gases that are usually present in a molten weld metal.

Blowholes

If the dissolved gases are present in amounts greater than their solubility limits, the excess is forced out of the solution in the form of a bubble or gas pockets as the metal solidifies. If the cavity is filled with paint it may cause bubbling due to air entrapment, so therefore it should be filled with a filling compound. Advice should be sought by the QC engineer on issues regarding steel defects and rectification.

CHAPTER TWO
Surface preparation

Importance of Surface Preparation

Of all the aspects of a protective coatings project, none are more critical to the success of the project than surface preparation. The best coating system can fail over an improperly cleaned and prepared surface. Even weak coating systems work better over properly cleaned and prepared substrates.

Management personnel often make choices when planning protective coatings projects that are based on "first cost" rather than "cost per metre square per year of service life"; this is done in the interest of meeting the annual budget. Such choices often involve the use of surface tolerant primers over marginally prepared substrates.

Although this choice satisfies the needs of meeting the annual budget, it ignores the fact that it will increase the forthcoming year's budget during the maintenance phase of any coating system applied over a marginally prepared substrate. Surface preparation is often thought of as that phase of a protective coatings project where mill scale, rust and old coatings are removed by some form of abrasive blasting or power tool cleaning.

Although this is critically important, there is an equally critical phase that precedes these removal methods. That phase is cleaning, which is necessary to remove any form of contaminant that is on the substrate prior to using any other removal method. All of the NACE and SSPC surface preparation standards begin by requiring the surface to be cleaned in accordance with the SSPC-SP1 "Solvent Cleaning Standard".

This does not mean the cleaning must be done with hydrocarbon-based solvents such as MEK or Xylene. Water is a solvent and is becoming increasingly popular as a cleaning method in conjunction with water based, biodegradable cleaners with some form of pressure cleaning. It is easy to see contaminants such as oil, grease, dirt, dust. It is also easy to see and test when they are properly removed. That is not a problem, as the real problem is with non-visual contaminants particularly water-soluble contaminants such as chlorides, sulphates and ferrous ions.

Heavily Corroded H-Beam and Bracing

These contaminants interfere with the ability of protective coatings to properly wet the substrate and achieve either a mechanical or chemical bond to the substrate. Unless they are properly removed, coatings applied over these contaminants will fail; the only questions are when they will fail and the method in which they will fail. Chloride and sulphates are the two most common water-soluble contaminants found on both new and previously coated substrates.

Any structure that serves in or is near a salt water body will have some level of chloride deposited on its surface either by the seawater itself or by prevailing winds blowing from the sea to the land-based structure. Structures that are in an industrial environment such as a refinery, chemical processing plant or fertilizer plant can be contaminated by sulphates that are carried downwind from that part of the plant process from which the sulphates are liberated.

The fact that chlorides and sulphates are normally non-visual (except in very high concentrations) is what makes them particularly destructive to properly applied coatings once these coatings have been placed into service. The dry, powdery chloride or sulphate deposit lies on the substrate and interferes with the wetting and adhesion of the protective coating.

That is bad enough, but the real problem comes when moisture works its way through the protective coating system as all coatings are semi-permeable to some degree and dissolves the powdery deposit. Once that powdery deposit becomes solubilized by moisture, it begins to grow in volume, by attracting more moisture through the film by osmosis creating pressure in the form of an osmotic blister. In addition to the dis-bonding pressure of osmotic blistering, soluble salts such as chlorides and sulphates decrease the life of a protective coating system by increasing the rate of corrosion under the coating film.

The important part of that corrosion cell from the standpoint of soluble salt contamination is the electrolyte. Water is a good conductor of electrical current, which also is an essential part of the corrosion cell. Pure water such as distilled or deionized water is a relatively poor conductor of electrical current, but contaminated water, particularly chloride contaminated water, is a very good conductor of electrical current. Therefore, when it solubilizes the powdery deposit of the chloride, it increases the conductivity, accelerating the activity of the corrosion cell.

Although there is no universally accepted level of what is allowable in the amount of chlorides and sulphates under protective coating systems, numerous papers have been published that demonstrate the effects of various levels of these water-soluble contaminants. Tests are available to measure the amount of water-soluble contaminants on a substrate. Standards widely used for the detection of water-soluble salts in the industry are ISO 8502–6/9. The cleaner the substrate being treated, the longer the life expectancy of the coating system.

Surface Preparation

Surface Preparation Comes in a Variety of Different Methods:

- Solvent Cleaning
- Water Washing
- High Pressure Water Jetting
- Shroud Blasting
- Slurry Blasting
- Steam Cleaning
- Abrasive Blast Cleaning
- Wet Blast Cleaning
- Hand Wire Brushing
- Power Wire Brushing
- Needle Gun
- Scaling Hammer

- Grinder and Sander
- Hand Scraper
- Chipping Hammer
- Sand Paper.

Solvent Cleaning

Solvent cleaning is a method for removing all visible oil, grease, soil, drawing and cutting compounds and other soluble contaminants from steel surfaces. Solvent cleaning is intended to be used prior to the application of paint and in connection with surface preparation methods specified for the removal of rust, mill scale or paint. SSPC-SP-1 is the only commonly used standard that formally governs solvent cleaning to remove oil, grease, dirt, drawing compounds and similar organic compounds. It defines a variety of pre-cleaning methods including:

- Solvent Wipe with Cloth or Rag
- Immersion of the Substrate in Solvent
- Solvent Spray
- Vapour Degreasing
- Stream Cleaning
- Emulsion Cleaning
- Chemical Paint Stripping
- Alkaline Cleaners.

A Variety of Solvent Cleaning Materials May Be Encountered:

Organic solvents

Kerosene, turpentine, naptha, mineral spirits, toluol, xylol, clean the metal by dissolving and diluting the oil and grease contamination on the surface. Some organic solvents used for solvent cleaning may be considered hazardous to health and are generally likely to be a fire risk. In particular toluol is subject to health and safety restrictions in many countries. Inorganic materials such as chlorides, sulphides, weld flux and mill scale are not removed by organic solvents.

The last wash or rinse should be made using clean solvent to remove the slight film of oil or grease that may be left on the surface. This film if left in place may interfere with the adhesion of the coating to the surface. Some solvents will also dissolve some paint films so they can be removed from the surface. Non-convertible coatings like chlorinated rubber and vinyl are most likely to be softened or removed by solvent washing. Petroleum-based mineral spirits with a minimum flash point of 38°C may be used as a general purpose solvent under normal conditions.

In hot weather 26°C to 32°C, high flash point mineral spirits with a minimum flash point of 50°C should be used. Most solvents are potentially hazardous and may be ingested in breathing air by applicators, helpers or others when solvent cleaning is performed. Working spaces should be monitored for solvent fumes, and concentrations in air breathed by workers should be below limits defined by state, or local regulations for threshold limit values (TLV).

Appropriate fresh air masks should be used in confined spaces and when the safe concentration may be exceeded. The fresh air supply should be free of carbon monoxide or other contaminants from other sources such as engine exhausts. The concentration of solvents in the air should not exceed the lower limit of flammability known as the lower explosive limit (LEL) or a fire or explosion could occur.

Such concentrations are most likely to occur in confined spaces such as tanks, pipes or vessels. In general, cost considerations and regulations restricting the use of organic solvents have become so stringent as to discourage the use of these materials except in highly controlled special situations.

Alkaline cleaners

Alkaline cleaners such as tri-sodium phosphate and sodium hydroxide saponify most oils and greases and their surface-active components wash away other contaminants. These cleaners may also saponify certain coating vehicles. Often alkaline cleaners are combined with surfactants (wetting agents), inhibitors, and detergents to form proprietary products and should be used accordingly to the manufacturer's recommendations. These products are often used at elevated temperatures.

In the cleaning process, a slightly soapy film may be left on the surface. This film must be removed, usually with a high-pressure hot water rinse. Residues that remain on the surface will affect the adhesion of the coating to the steel surface. A pH paper test could be made to determine the effectiveness of the wash. In general, the pH of the washed surface should not exceed the pH of the wash water.

Acid cleaners

Acid cleaners are usually composed of fairly strong acids, such as phosphoric acid with small quantities of surfactants, water-miscible solvents, and organic wetting and emulsifying agents. Acidic cleaners remove contamination by chemical attack and by dissolving the reaction products. They may be used to remove corrosion products and for other special purposes. Disposal of alkaline or acidic cleaners is often a problem and collection of the waste or used material is essential. Used materials cannot be washed into the normal drainage system or allowed to run off into the general drainage system.

Detergents

Increasingly strict regulations regarding the use of organic solvents and the safety considerations in using alkaline or acidic cleaners have led to an increased use of detergents, especially the biodegradable types to remove oil and grease, and other similar contaminants from the surface. Generally these cleaners are composed of buffering salts, dispersants, soaps and inhibitors, and they function by wetting, emulsifying, dispersing and solubilizing the contaminants, which can be washed away using water (usually hot) or steam. They are frequently used at temperatures ranging from 65°C to 100°C.

Emulsion Cleaners

Emulsion cleaners typically are proprietary products and should be used in accordance with the manufacturer's instructions. These cleaners may contain oil soluble soaps or emulsifying agents, buffering salts, dispersants, and inhibitors, along with kerosene or some types of mineral spirits. Emulsion cleaners are generally sprayed onto the surface to be cleaned and they function by wetting, emulsifying, dispersing, and solubilizing the contaminants.

Generally the emulsion cleaners on the surface leave a thin film of oil which must be washed away with hot water, steam, solvents, detergents, or some type of alkaline cleaning agents. Water is a common, if not powerful, solvent and may be used to good effect in cleaning a surface. In the use of any cleaning method appropriate safety precautions must be taken. Also it is important to thoroughly rinse surfaces, particularly when alkaline or acidic materials are used to minimize the amount of contamination remaining and to remove residues of cleaning materials that could adversely affect subsequent coating performance.

Water Washing

When steel arrives on site surface degreasing and water washing is usually required to remove surface contaminants such as soluble salts (sodium chloride) oil, grease and cutting compounds. Pressures range from 1,000 psi to 5,000 psi and should not be confused with high pressure washing or water jetting UHP. It is important that fresh water is used when preparing surfaces for coatings due to the fact that seawater (salt water) contains sodium chloride and if used it will have a detrimental effect on the performance of the coating system being applied. If salt water had been used due to work location, and availability of fresh water, then the final wash MUST be with fresh water.

Water Washing

Salt water will also have a massive impact on the water washing machine's internal workings and may cause major damage. If seawater is used the machine must be flushed thoroughly to remove salts (chlorides) after each use. Always seek advice from the pump manufacturer as it is he who will be able to give the best advice about the pump's performance and any guarantees/ assurances regarding his pump. The use of detergents may be needed to remove impurities from a heavily contaminated surface such as oil or grease.

The detergent can be applied by spray or by brush; however, if the surface is heavily contaminated it may be necessary to scrub the surface with stiff bristle brushes. PPE, Personal Protective Equipment, is necessary even when using pressures of 1,000 psi to 5,000 psi. This will consist of a waterproof suit which is best breathable to allow the operator to work comfortably. A waterproof hood is also worn to allow maximum visibility when washing.

Water Washing Advantages:

- Easy to operate.
- Remove soluble contaminants such as Sodium Chlorides.
- Removes loose coatings.

Detergents can be added to remove oil and grease

- Pressures range from 1,000 psi to 5,000 psi.
- No dust is created.

Water Washing Disadvantages:

- High pressures are used which can be dangerous.
- Fresh water is required.
- Electrical equipment must be thoroughly encapsulated.
- Contaminated water must be properly disposed of.

High Pressure Water Jetting

High Pressure Water Jetting is a method of surface preparation that uses water at very high pressures projected onto a surface. No abrasive is introduced into the stream of water as water jetting relies entirely on the energy of water striking the surface. All coatings and contaminants have a removal threshold energy, the mode of which is usually sheer rather than tensile. If the water jetting forces exceeds this threshold, the coating and contaminants will be removed.

High Pressure Water Jetting

If these contaminants are not removed and blast cleaning commences, the contaminants will be distributed over the substrate inevitably leading to premature coating failure. It is important when water jetting that fresh water is used and not seawater/salt water. The salt water can damage water blasting equipment and push water soluble contaminants into the substrate.

If seawater/salt water is used it is imperative the surface is fresh water washed to remove these contaminants or premature coating failure will occur. Organic solvents and degreasing agents should be used to remove oil and grease with fresh water washing being performed after the use of such operations. SSPC-SP1 is a standard widely used in the removal of such contaminants.

After the removal of surface contaminants such as oil and grease by solvent cleaning (SSPC-SP1) water jetting is necessary to remove surface residues and water soluble contamination. If the salts are not removed by water jetting then osmosis will take place leading to premature coating failure.

Water Jetting requires at least two people, one to operate the pump and the other operates the lance. The UHP pump creates pressure which drives the water through the hoses then out of a handheld lance. The lance is available with multiple nozzles and various lengths. Different types of coatings require different flow rates and pressures for removal. The pressure is regulated by the pump and the flow rate is determined by the size of the tips in the nozzles.

There needs to be good coordination between the pump man and the operator using the lance. The pump operator should bring up the pressure slowly as the operator on the lance requires a firm footing with such high pressures. Increasing the pressure suddenly could throw the operator off his feet resulting in serious injury. Specialist safety equipment and training is essential when operating high pressure equipment.

Ultra-High Pressure Pump

The risk of water injection and operator fatigue are the two main risk factors associated with UHP. If injected by water micro-organisms can enter through the wound and spread into the body causing a fatality. The removal of a high proportion of soluble salts from the steel surface is seen to be the major advantage of water jetting. Rust and coatings are sheared and lifted from the surface relatively easily compared with other blasting methods.

The most commonly used nozzle assembly is the rotating head as it increases the size of the blast pattern and increases the cutting action. Productive cleaning requires the nozzle to be held 6mm to 13mm from the surface. Soluble salts are removed more effectively because the steel profile generally remains unchanged; however, water jetting will only reveal the existing profile, it will not create one, and is primarily used in maintenance projects.

By injecting small amounts of abrasive into the water stream it can produce a rougher surface profile. Water Jetting is an extremely versatile and effective method of removing paint and thick coatings, soluble salts and other contaminants from the surface. It is environmentally friendly and excels around sensitive equipment and air intakes where grit could get into moving parts, possibly shutting down a compressor or turbine.

Water Jetting surfaces are cleaner than those using expendable abrasives; in addition, soluble contaminants (sodium chlorides) are removed, which results in better adhesion of the coating and longer life expectancy. With water jetting no abrasive is forced into the surface profile causing discolouration, unlike iron silicates or expendable abrasives.

Water Jetting Advantages:

- Removes oil and grease.
- Removes soluble contaminants such as chlorides.
- No dust is created during this process.
- Good at removing thick and robust coatings.

Water Jetting Disadvantages:

- Availability of clean fresh water.
- Does not create a surface profile, only reveals the existing one.
- Water jetting is difficult in enclosed and small areas.
- Visibility is poor in enclosed areas and confined spaces.
- Prepared surface quickly flash rusts.

Pressure Definitions:

- **Low Pressure Water Cleaning**: Operates at Pressures below 34 MPa (5000 psi)
- **High Pressure Water Cleaning**: Operates at Pressures between 34 to 70 MPa (5,000psi to 10,000 psi)
- **High Pressure Water Jetting**: Operates at Pressures between 70 to 210 MPa (10,000 psi to 30,000 psi)
- **Ultra-High Pressure Water Jetting**: Operates at Pressures above 210 MPa (30,000 psi).

Low Pressure Water Cleaning

Low pressure water cleaning used for surface preparation is primarily a washing technique. At pressures below 34 MPa (5,000 psi) water removes soluble contamination and some loosely adherent surface contaminants. It will reliably remove chalking of aged coatings, leaving the coating surface intact. Low pressure water cleaning is often used to wash the underside of ships in dry dock removing marine foulants and growth.

High Pressure Water Cleaning

High pressure water cleaning is commonly used for surface preparation of concrete surfaces prior to coating application. When used for surface preparation for coatings, the production rate is generally low. In addition, only loose contamination can be successfully removed.

High Pressure Water Jetting

High pressure water jetting equipment is seldom used for surface preparation for coatings. The cleaning effect is not better than equipment operating at lower pressure and the production rate is not cost effective.

Ultra-High Pressure Water Jetting

This method uses water at very high pressure 210 MPa (30,000 psi) and above up to 340 MPa (50,000 psi). Because of the high pressures required, safe practice demands great care in controlling the water jet nozzles, since a person struck by high velocity water at short range could be seriously injured. Most UHP WJ equipment operates with a rotating nozzle and dual water streams. The highly effective nozzle design produces an effective cleaning pattern while using relatively little water, perhaps no more than 8 litres per minute.

The nozzle must be held close to the surface being cleaned, as the cleaning effect decreases rapidly at some distance from the nozzle. The water stream has almost no effect on the surface if the nozzle to surface distance is increased to more than 50 cm. The most effective cleaning effect is achieved at a spray distance of approximately 5 cm.

Water used at this pressure removes most contaminants such as salts, grease, dirt and rust scale. It will not produce a surface profile but is capable of restoring the original profile wherever it previously existed.

One incidental benefit of the very high pressures used is a heating effect on the surface being blasted. In the case of steel, this heat has the effect of limiting the deterioration through rusting, and the surface remains relatively clean although some gingering may occur.

Joint Surface Preparation, Standard Water Jet Cleaning of Metals NACE/SSPC- SP WJ 1,2,3,4

Description of the surface appearance after cleaning

- **WJ-1 – Clean to Bare Substrate** (WJ-1): A metal surface after clean to bare substrate, when viewed without magnification, shall have a matte (dull, mottled) finish and shall be free of all visible oil, grease, dirt, rust and other corrosion products, previous coatings, mill scale and foreign matter.
- **WJ-2 – Very Thorough Cleaning** (WJ-2): A metal surface after very thorough cleaning, when viewed without magnification, shall have a matte (dull, mottled) finish and shall be free of all visible oil, grease, dirt, rust, and other corrosion products except for randomly dispersed stains of rust and other corrosion products, tightly adherent thin coatings, and other tightly adherent foreign matter.

  The staining or tightly adherent matter shall be limited to no more than 5% of each unit area of surface and may consist of randomly dispersed stains of rust and other corrosion products or previously applied coating, tightly adherent thin coatings, and other tightly adherent foreign matter.
- **WJ-3 – Thorough Cleaning (WJ-3)**: A metal surface after thorough cleaning, when viewed without magnification, shall have a matte (dull, mottled) finish and shall be free of all visible oil, grease, dirt, rust, and other corrosion products

except for randomly dispersed stains of rust and other corrosion products, tightly adherent thin coatings, and other tightly adherent foreign matter.

The staining or tightly adherent matter shall be limited to no more than 33% of each unit area of surface and may consist of randomly dispersed stains of rust and other corrosion products or previously applied coating, tightly adherent thin coatings, and other tightly adherent foreign matter.

- **WJ-4 – Light Cleaning** (WJ-4): A metal surface after Light Cleaning, when viewed without magnification, shall be free of all visible oil, grease, dirt, dust, loose mill scale, loose rust and other corrosion products, and loose coating. Any residual material shall be tightly adhered to the metal substrate and may consist of randomly dispersed stains of rust and other corrosion products or previously applied coating, tightly adherent thin coatings, and other tightly adherent foreign matter.

ISO 8501-4:2006 Water Jetting Standard

BS EN ISO 8501-4:2006 which is titled "Preparation of Steel Substrates before Application of Paints and Related Products – Visual assessment of surface cleanliness – Part 4: Initial surface conditions, preparation grades and flash rust grades in connection with high-pressure water jetting".

The standard specifies a series of preparation grades for steel surfaces after removal/ partial removal of water-soluble contaminants, rust, previous paint coatings and foreign matter by high-pressure water jetting. The various grades are defined by written descriptions together with photographs that are representative examples within the tolerances for each grade as described in words. The standard contains 23 representative photographic examples in the following categories:

ISO 8501-4:2006

- Five initial surface conditions, three of them applicable to degraded paint coatings and two of them to damaged pre-fabrication (shop) primer coatings.
- Three preparation grades for each initial surface condition, after partial or full removal of previous paint coatings by high-pressure water jetting.
- Three flash rust grades after pre-treatment by high pressure water jetting.

Shroud Blasting

One of the advantages of using water in a blast cleaning operation is the reduction of the dust hazard. This aspect is particularly important when blast cleaning takes place in a relatively public environment, as when city buildings are blast cleaned. For this reason, equipment was developed to provide a shroud of water around a normal flow of an abrasive carried in compressed air.

Other advantages of this modified dry blasting system include the ability to remove soluble contaminants from the surface. Hazardous components of the waste, including dust emissions, are much reduced. Up to 75% of dust cannot escape from the water stream and is in theory unable to enter the immediate surrounding environment.

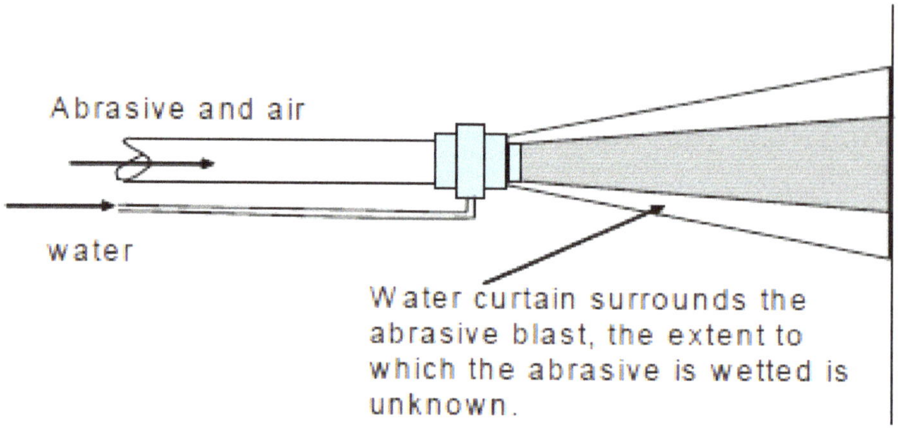

Example of Shroud Blasting

Disadvantages include the problem of removing the used abrasive in its wet form and the necessity of using an inhibitor in the water to prevent surface rusting when used to clean steel. The water ring attachment is connected to a low pressure water source and the flow of the water is controlled by a small valve on the attachment.

In use the water stream wets the abrasive stream outside and in front of the nozzle discharge, which helps keep dust to a minimum. This cleaning process may be used where dust would be objectionable. The surface profile is similar to that achieved with dry abrasive blasting, but of course the surface is wet after blasting. Blast pressures are much the same as for dry abrasive blasting up to 100 psi (7 bar) and production rates are similar. Clean up time is probably more because of the difficulty of removing wet abrasive sludge.

Slurry Blasting with Abrasive

In this method the abrasive and water are mixed together at or near the blast pot with constant agitation to form a slurry. The slurry then is pumped through a single hose to the blast nozzle. Because the abrasive water mix is pumped as slurry the pressure can be easily controlled. This means the cutting effect of the abrasive can be increased or reduced at will, and in fine increments to allow special effects such as removal of only the top coat or feathering back the coating edges.

Steam Cleaning

Steam cleaning is not widely used in the protective coatings industry as much as high pressure washing or water jetting. Steam cleaning is used to remove contaminants such as oil and grease deposits on steel substrates as well as coated surfaces. It is not only used in the coatings industry as councils use such methods to remove graffiti from concrete and brickwork structures.

High temperatures above 100°C and high pressures are used with steam cleaning so the correct training is required as the process can be dangerous in untrained hands. The use of hot pressurized steam is a very effective method of cleaning surfaces. Care must be taken not to damage the coatings being cleaned as emulsions and waterborne coatings can sustain damage when using temperatures of above 100 °C.

Steam Cleaning

As with high pressure washing and water jetting, Steam cleaning does not create a surface profile, it only reveals the existing profile. With high temperatures generated during steam cleaning, the correct Personal Protective Equipment (PPE) must be worn. This includes a breathable waterproof suit which must withstand high temperatures, a waterproof hood, waterproof boots, gloves.

Abrasive Blast Cleaning

In the protective coatings industry, the preferred method of preparing steel substrates is by abrasive blast cleaning. This operation propels a stream of abrasive material against a surface under high pressure. The blast cleaning process was first patented by Benjamin Chew Tilghman on 18 October 1870 in Pennsylvania, USA. Tilghman had seen the effect of sand blown against a glass window while in the desert and that was the basis for his sand blasting invention.

Benjamin Chew Tilghman

The process for blast cleaning is quite straight forward and has never changed much over the years. Blast cleaning is used all over the globe on offshore structures, shipyards, construction sites and bridges. The process removes surface contaminants such as rust, mill scale and paint coatings; it will also roughen the surface and increase the surface area. As the surface is roughened a surface profile is created and this will enhance mechanical adhesion.

"The primary objective of surface preparation is to provide maximum coating adhesion."

Blast Cleaning Equipment

Components/Accessories of an Air Pressure Blasting System are:

- Pressurized Blast Pot
- Metering Valve
- Blasting Hoses
- Lighting
- Compressor
- Blasting Nozzles
- Hypodermic Needle Gauge
- Couplings
- Blast Suit
- Hand Protection
- Air fed Blast Helmet
- Water and Oil Separator
- After-Cooler
- Fresh Air Filter
- Air Receiver Tank
- Dead Man's Handle
- Blasting Sieve.

Pressurized Blast Pot

Compressed air is distributed to both the blasting hose and the blast pot by a compressor; this may be either diesel driven or electric. This is to ensure that the pressure is the same in the blast pot as in the hose. The abrasive will then run at an even flow through the metering valve and into the jet of air. When the abrasives are used up, the air supply is shut off both to the blasting hose and to the blast pot.

The "mushroom" or pop-up valve at the top of the blast pot is opened and the air released. In this way the pot is depressurized more quickly than when the air is released through the blasting nozzle. The pot is now pressure-free and fresh abrasives can be replenished.

Very few changes have actually been made to the blasting unit during more than 100 years since Benjamin Chew designed it; however, some developments have taken place.

Today all blasting units have a cylindrical shape with the bottom part being cone shaped and ends in a pipe stub where the metering valve is connected. The cone wall is so steep that the abrasive can flow down under its own weight.

Full Size Blast Pot

The top of the unit is also conically shaped like the bottom part with an opening in the middle for refilling abrasives. Around the opening on the underside of the top plate, a thick rubber packing is fitted; this is often shaped like an O-ring. When the tank is pressurized and the blast pot put into action, the air pressure lifts the "mushroom" pop-up valve to fit closely against the rubber packing. If the metering valve is open, abrasives will now flow into the blast hose.

Pop-up Valve or "Mushroom"

When the abrasive/air mixture flows evenly from the nozzle, the blast cleaning process can start. The pop-up valves may vary in shape and design; they are conical and may be more or less pointed at the top or shaped like a "mushroom". Today practically all cones have a core of steel coated with a hard-wearing rubber compound. When the refilling of abrasive takes place from a bag or hopper, it is common practice to fill it up right below the pop-up valve.

Mid Sized Blast Pot

The abrasive should not cover up the entire pop-up valve as air will be lost and the pot will not be fully pressurized. An uncoated metal valve with a pointed top has commonly been used in blast pots with continuous filling. In such units the abrasives are stored in a hopper which is located above the blasting units. When the operator stops blasting and the blast pot is depressurized, the abrasives automatically flow from the hopper to the blast pots. When the operator starts up again, the pop-up valve must force it way through the abrasive in order to fit tightly against the packing.

Metering Valve (Miser Valve)

A selection can be made between several types of abrasive metering valves. If a certain valve is chosen, this is usually done on the basis of the type of abrasive to be used. However, the choice also depends on whether the valve is to open and close automatically. Whether the valve is automatic or not, it does not depend on the abrasive but on the equipment.

If the blast cleaning is carried out with copper slag abrasive, a miser valve is often mounted on the blast unit. The housing around the miser valve contains two plates with an orifice hole in each of the plates. The plates are mounted at the end of the stem, one plate is fixed whilst the other plate is movable with the stem. The stem extends beyond the housing, and beyond the housing a regulating handle is fitted.

If the handle is moved, one of the plates is moved as well. When the handle is in mid position the orifice holes in the plates fully overlap; in this case a full opening is formed. The more the handle is taken to one of the sides the less overlap of the orifice is produced. The handle and the orifice should be adjusted in such a way that the valve is fully opened in mid position and completely shut with the handle in an extreme position.

On miser type valves a guide pin is often mounted on the regulating handle, the guide pin can be marked with a scale showing the position of the orifices relative to each other. In addition, the handle can be screwed fast to the guide in such a way that the adjusted setting does not change.

In units for blasting with metallic abrasives, other types of metering valves are used rather than the miser types. They are usually called steel grit valves as many people call all metallic abrasives "steel grit" especially the sharp edged types. The steel grit valves can be distinguished from the miser valves by the way the volume of abrasive is regulated and how the abrasives flow into the air jet.

Miser Valve

Two of the most common methods of regulation are as follows:

1 – Metering valve with a rubber case for vertical passage of metallic abrasive. This valve is mounted in such a way that the abrasives flow directly from the blast pot through the opening of the rubber case. By means of a plunger fitted to a spindle with a regulating knob at the end of the opening, the rubber case can be changed. If the spindle is screwed in, the plunger will squeeze the rubber case together. Thereby both the opening and the volume of abrasives that can pass through are reduced.

2 – Abrasive control valve with a ball shaped valve for vertical passage of metallic abrasives: The steel grit valve is regulated by adjusting the opening between a ball shaped valve and a valve seating. A small opening results in only a small volume of abrasives passing through. In this steel grit valve, the parts which are in contact with the abrasives should be made of hardened, highly abrasive resistant steel or be coated with other abrasive resistant materials.

The abrasives flow through the miser valve and into the blasting hose at the optimum angle of 30% as that is as parallel to the air jet as possible without the angle of tilt becoming so small that the abrasives stop running. Thereby the abrasives pick up speed more easily from the compressed air. This benefit is not obtained if the abrasives flow at right angles into the air jet.

When the operator stops blasting, by means of the dead-man, thereby the blasting hose is emptied of abrasives. When blast cleaning is resumed, it is not necessary to put into motion "dead" abrasives which would otherwise have been present in the hose. After the blasting has finished the automatic metering valve can be closed, but the compressed air inlet held open. It is then possible to use pure compressed air for blowing away abrasive and dust.

The volume of abrasives necessary for accomplishing this depends on the:

- Constructional Design
- Condition of Surface – Degree of Rust
- Abrasive – Type, Grain Size
- Dust Formation – Visibility, Lighting Conditions
- Compressor Pressure
- Hose Length
- Nozzle Type and Size
- Operator.

A few indications of incorrect abrasive/air mixture are high noise or a whistling sound when the abrasive supply is too small, and if it's too big the mixture is ejected unevenly and intermittently from the nozzle. At the same time there is a jerking action inside the blast hose.

Blasting Hoses

They are two types of Blast Hose:

- **Four Ply,** for use when the hose is subject to external abuse or when there is danger of the operator pulling it at right angles
- **Two Ply,** a lighter hose that is preferred by some blasters because of its greater flexibility.

The objective of the blast hose is to carry abrasives at high pressure on the substrate. The abrasive carried through the blast hose could be metallic, mineral or synthetic. The blast hose should be of good quality and made from the correct grade of material; if not a rupture could occur, exposing abrasive and pressurized air, potentially injuring personnel.

The safe working pressure for the blasting hose is between 8 bar and 12 bar; the pressure is identified on the side of the blast hose. The hose should be checked for damage such as punctures, cracks and blisters before and after each shift. The internal of the hose is impregnated with carbon (anti-static inner wall) as this eliminates the risk of static shock. Blasting hoses are generally manufactured in 10, 20, 30 metre lengths.

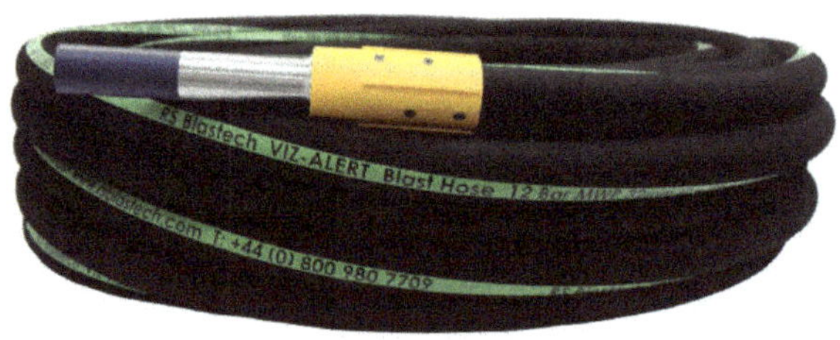

Blasting Hose

It is always best to get the blasting pot as near to the workplace as possible to minimize hose length. It is not uncommon to have blast hoses over 300 ft away from the blasting set-up. The longer the length of the blast hose, the less pressure at the blasting nozzle. The diameter of the blasting hose is important because of the friction loss measured in pressure drop of conveying abrasives. Only good quality blasting hoses should be purchased with a robust outer layer required.

The blast hose should be kept as straight as possible whilst blasting activities are ongoing. If the hose is bent or kinked, it will restrict the abrasive and airflow that could lead to a rupture of the blast hose; also the hose will become worn more quickly. When the hose is not in use it should be stored in a clean, dry environment away from exposed sunlight. The hose should be treated with care and respect as this way it will have a longer working life.

Suggestions for Blast Hoses:

- Make sure the blast hose is as straight as possible when in use
- Ensure no equipment or consumables are placed on top of the blast hose
- Make sure there are no kinks or damages to the blast hose as these will be weak points
- Blast hoses and blasting consumables should be kept in a dry storage area.

Air Compressors

In the blast cleaning operation, the air compressor must be capable of delivering sufficient air volume to maintain the required pressure at the nozzle. Related equipment including the following must be of sufficient size and type to match the compressor output:

- Air and Abrasive Hoses
- Hose Couplings
- Blast Nozzles.

The compressed air must be free from contaminants including oil and water. Firstly, since air supplied breathing equipment is required when abrasive blast operations are being performed, it is critical that the breathing air be fresh and pure. This is the responsibility of the contractor. Secondly, it is important that the blast air be clean to ensure that the blast cleaning operation does not add contaminants to the surface being cleaned.

The blasting contractor is responsible for complying with all requirements of the specification, such as the size of the hose, type of nozzle and air volume, if these are detailed in the specifications. The coating inspector may check each of the requirements specified to ensure compliance by the contractor. Compressed air is a common source of power for blasting machinery, paint spray equipment, power tools.

Typical Air Compressor

To produce quantities of compressed air it is necessary to use a compressor. Normally driven by a diesel motor although electric compressors are becoming more common in fixed workshops. A compressor draws in atmospheric air, pressurizes it, and feeds the air into a pressure vessel also known as a receiver. The air is then held in the receiver until demanded by the equipment in use. The production of compressed air presents two problems to the surface preparation process; these are:

- Any change in atmospheric pressure may result in the release of water vapour from the air.
- Because compressed air in the receiver is stored by pressurizing an oil reservoir, there is a possibility of oil vapour being retained by the air as it is released.

Both of these factors require that adequate vapour traps are fitted to blast cleaning equipment to remove the contaminating oil and water. Compressors are rated for both air pressure and capacity.

Air pressure

Air pressure is measured in Pascals (Pa) and pounds per square inch (psi) or bar. Air pressure is normally set a little above the intended working pressure, normally a maximum of 100 psi – 7 bar for portable compressors. This pressure, if successfully maintained, produces efficient blast cleaning. Newer compressors can maintain pressures up to 150 psi – 10.5 bar. Air pressure used should be no more than safety considerations and regulations allow. This figure may vary in different countries.

Adequate pressure is critical to an effective blast process. If the abrasive is not propelled at sufficient pressure, extra time is required to accomplish the job. The compressor pressure control should be set to compensate for pressure lost in the blast hose. Pressure at the nozzle can be measured using a hypodermic needle gauge inserted just before the blast nozzle.

Capacity

Capacity is measured in cubic feet per minute (cfm). The capacity of a compressor will determine the quality of air it can deliver at its working pressure. For blast cleaning purposes it is better to have a large capacity compressor working below its maximum level than a smaller compressor working at or near its maximum level. The compressor selected should be able to supply more air than required to allow for reserve capacity for peak loads or for the addition of equipment.

For blast cleaning using an 11mm diameter nozzle operating from one blast pot it is generally recommended that a compressor with a minimum capacity of 350 cfm be used. For extended use a compressor with greater capacity, perhaps 600 cfm, will operate with less strain and greater efficiency. The same compressor used for blasting is often used to drive air operated spray guns and other equipment. No other equipment requires such large quantities of compressed air as the blast cleaning operation.

It is estimated that with one 9.5mm nozzle operating there will be a 5psi drop in pressure for each 50ft of hose in use. This pressure drop will be compounded according to the number of operating nozzles, the nozzle sizes, and the length of hose being used. Undersized compressors create inefficiencies that directly impact cost and schedules; this may lead to:

- Time wasted waiting for the compressor to build up to the required pressure when its capacity is not great enough.
- Undue strain placed on the compressor during peak loads.
- Loss of time caused by air-starved equipment operating inefficiently.
- Inability to add new tools to the system.
- Greater possibilities of breakdowns or shutdowns.
- Excessive running to supply the quantity of air needed.
- Excessive heat and condensation.

Another consideration is the size of compressor manifold valve. The compressor should be equipped with outlets of 38mm to 50mm to match the air supply hose.

Blasting Nozzles

Many types of blast cleaning nozzles dominate the protective coatings industry and the type of nozzle used on the job can have a big impact on production rates. Specialised nozzles are also available – these can be made from boron alloy, tungsten carbide and silicon carbide, and can either be short or angled. The abrasive is projected through the blast hose and nozzle onto the substate being treated. Nozzles comes in many forms and are made of quite robust material:

- **Straight Bore Nozzle** – Good for spot blasting.
- **Venturi Bore Nozzle** – Creates a wide blast pattern.
- **Double Venturi** – Increases the size of the blast pattern.
- **Wide Throat Nozzle** – Good for increased productivity.
- **Angle Nozzles** – Good for hard to reach locations.

Blasting Nozzles come in a Wide Range of Specialist Materials

- Boron Carbide
- Tungsten Carbide
- Aluminium
- Silicon Nitrate

Boron Alloy Nozzles

A very popular blasting nozzle and relatively cheap compared to other types of nozzles used in the industry. The life expectancy is said to between 40–60 hours depending on the type of abrasive is used. Due to the hardness of the material, even when aggressive abrasives are used the Boron nozzle has the following advantages:

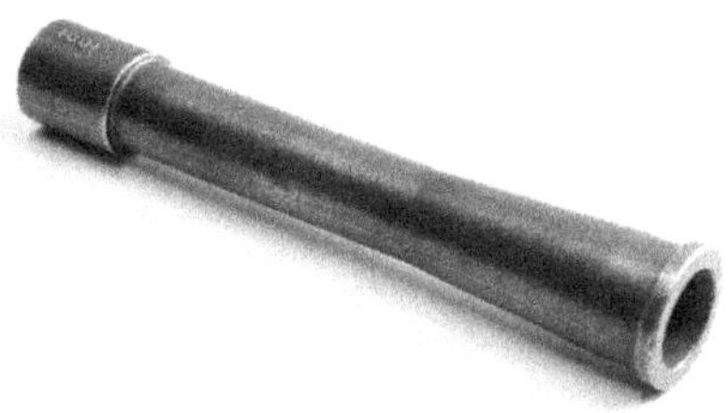

Boron Alloy Nozzle

- Long Working Life up to 25 times longer than some nozzles.
- Low Cost of the nozzle.
- High Production rates.

Tungsten Carbide Nozzles

Tungsten carbide nozzles are available as venturi type nozzles; long or short versions are available. Tungsten carbide is a hard, heavy material used in various industries due to its high abrasive resistance. These are amongst the most common type of nozzle used in the grit blasting industry as they are tough and best suited for use in a harsh working environment such as offshore. Tungsten nozzles work well with most abrasives, are lightweight and have a long wear life.

Tungsten Carbide Nozzle

Aluminium Nozzles

Aluminium nozzles are designed as a general-purpose nozzle with an excellent level of robustness and durability. The aluminium venturi provides superior performance greatly increasing the abrasive velocity while enhancing the cleaning speed. These are available in long or short lengths with the life expectancy of between 900–1200 hours depending on the type of abrasive used.

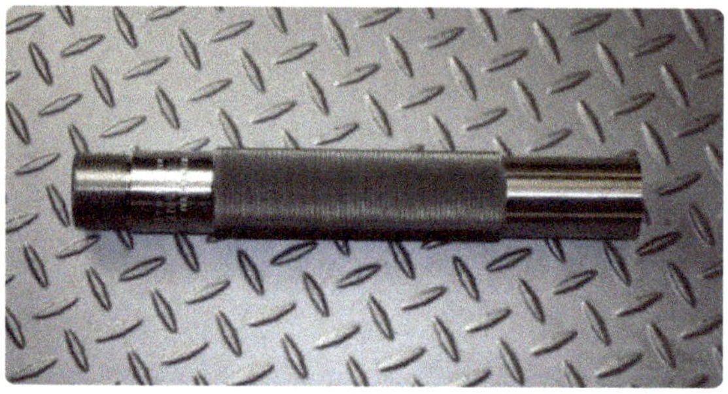

Aluminium Nozzle

Silicon Nitrate

Silicon Nitrate nozzles are tough and lightweight, offering excellent blast cleaning rates with consistent performance and economy over a long operating life. The nitrate jacket absorbs site impact and reduces operator fatigue when blasting over a long period of time. If handled correctly and not abused, a long service life is guaranteed. Nitrate nozzles work well with a number of blasting abrasives and increase production rates of up to 30% due to the size and strength of the nozzle.

Silicon Nitrate Nozzle

Nozzle Types

Straight Bore Nozzles

Straight bore blasting nozzles have been used in the industry for many years with this particular nozzle giving the full impact of the abrasive in the centre of the work piece.

The velocity of abrasive leaving the nozzle is approx 200 MPH with the impact of abrasive contacting the surface within a 6 inch diameter. The impact of the of the abrasive is only in the centre section only. Full impact is at the centre of the blast cleaned area with partial impact being around the edges.

Straight Bore Nozzle

Venturi Blasting Nozzles

Venturi shaped nozzles have been around for almost 50 years and with this particular design of nozzle a more uniform blast pattern is achieved. The venturi nozzle is very popular in the protective coatings industry and is amongst the blast cleaning operatives' most favoured nozzle.

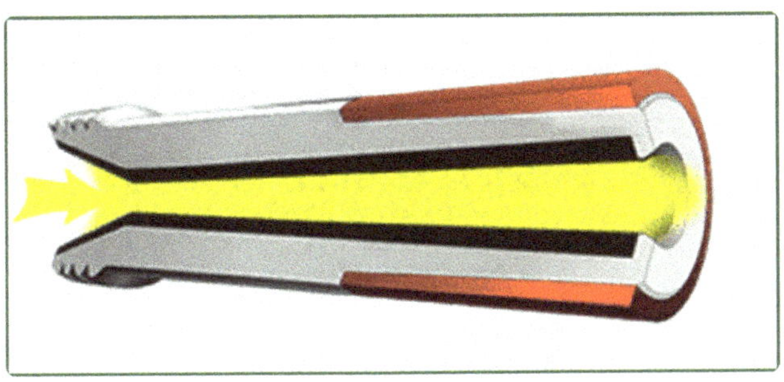

Venturi Blasting Nozzle

The full impact of the blast cleaning pattern is over 6 inches in diameter; there is no partial impact with the venturi shaped nozzle. The venturi nozzle is far more effective and quicker cleaning times are achieved over the use of a straight bore nozzle. The venturi nozzle has an open mouth, which allows a better flow of abrasive, resulting in quicker cleaning times, especially on large surfaces.

Double Venturi Blast Nozzle

Double venturi blast nozzles are designed to increase the velocity of the air and abrasive stream by drawing additional air through the unique induction holes, allowing a larger blast pattern. With the blasting pattern being much larger, higher production rates are also produced. These nozzles provide a medium abrasive cutting action, making them suitable for use with specialist abrasives like plastics. The blast pattern is approximately 35% larger than conventional venturi nozzles, along with a larger blast pattern and even uniform finish.

Double Venturi Nozzle

Wide Throat Blasting Nozzle

Wide throat blast cleaning nozzles enhance and increase production by up to 40% over a venturi blasting nozzle. These nozzles have a large entry throat and a large diverging exit bore and are sometimes called the "bazooka". Lower abrasive use is associated with these types of nozzles and can also be used at higher blasting pressures.

Wide Throat Nozzle

Angle Nozzles

These nozzles have been used with great success on blast cleaning projects and can reach locations where other types of nozzles cannot reach. These nozzles are lined with tungsten carbide and due to the 45 degree angle they tend to wear away considerably faster than traditional tungsten nozzles. These nozzles are ideal for tight and difficult locations where the blast cleaning operative cannot see due to the configuration of the work space. Angle nozzles are excellent for lattice steel work and small box sections that cannot be physically accessed.

Angle Nozzle

Wand Nozzle

The wand nozzle is used for extremely tight locations like in between girders and inaccessible spots. When necessary these wands can be extended to provide a longer reach. Wand nozzles can be used with most abrasives and only require a 20mm clearance in-between the work piece. Good surface preparation results can be achieved but cleaning rates are slow.

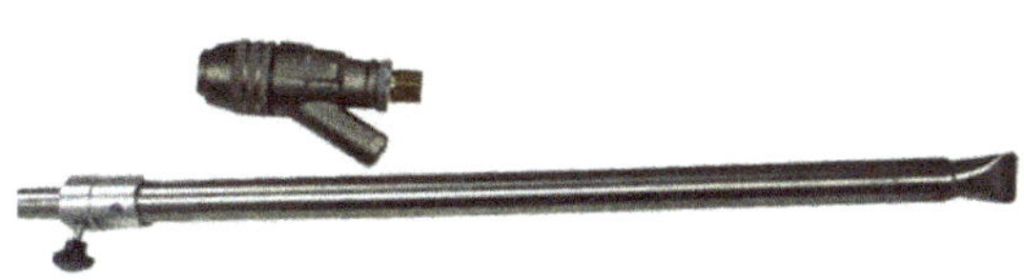

Wand Nozzle

Hypodermic Needle Gauge

The needle gauge is used to measure the air pressure at the nozzle. A hypodermic needle is inserted into the blast hose just behind the nozzle at a 45 degree angle while the blast unit is in operation. As soon as the needle penetrates the hose and meets the air stream, the pressure is registered on the gauge. Normal blasting pressures should be between 90 psi and 100 psi at the nozzle. The test is run with both air and abrasive flowing. If the pressure is less than expected or desired, the following items should be checked:

- Compressor size
- Size and length of air lines
- Couplings (for leaks, gaskets out of place)
- Size and type of nozzle
- Blast hose lining (for defects)
- Pipe on blast pot (for partial blockage)

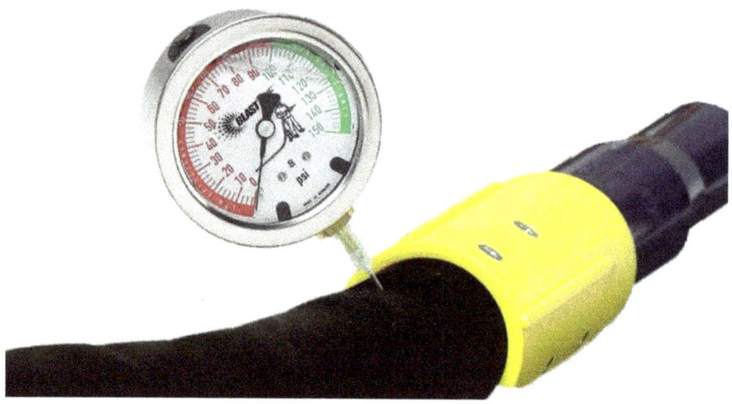

Hypodermic Needle Gauge

Blasting Couplings

The abrasive blast hose is connected to the blast pot with the use of a coupling. Clamps are used to keep the coupling in place on the blast hose; these are available in plastic, bronze and brass. The coupling is attached to the end of the blast hose by screws which should be of the correct size and tightly fitted. Care must be taken when attaching a coupling to a blast hose as the screws can penetrate the wall of the hose. Inside of the coupling is a gasket that should be of the correct diameter and should be checked for cracks or damages.

Blasting Couplings

Whip Checks

Safety wires or whip checks are an essential safety device which should be attached to the coupling and checked at regular intervals for damage. The safety wire is used as a back-up safety device in case the coupling is blown apart from the blast pot; with the safety wire being in place it prevents the hose from losing control and causing injuries. When repairing the blast hose due to rupture, the defective piece of hose is cut away with a new coupling re-attached; however, this should be conducted by experienced personnel.

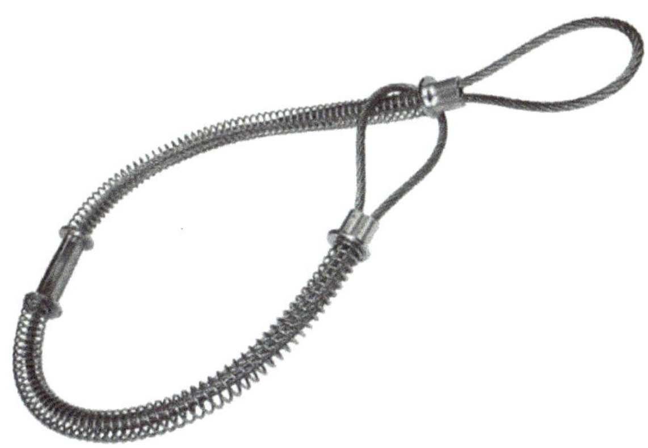

Whip Check

Personal Protective Equipment (PPE)

Blast Suit

Abrasive blast cleaning is a perilous task to the operator as high pressures are required to project abrasive particles onto the substrate at very high speed (450 MPH). The abrasive particles rebound back off the surface being prepared onto the blast cleaning operator. It is therefore essential that the correct PPE is worn when blast cleaning operations are ongoing, together with a fresh air supply as conditions are often dusty and dirty with the work location frequently being confined.

Blasting Suit

Dust and debris are created with the blast cleaning process as the abrasive breaks down as it impacts the substrate being prepared. The coatings and corrosion being removed while blasting will also be in the spent blast media creating a dust-laden atmosphere. The blast cleaning operator will be required to wear full body protection together with a fresh air supply to the helmet.

The suit comprises rugged leather panels that protect the arms and legs from rebounding abrasives and leather gauntlets that protect the arms and wrists. The blasting helmets are manufactured in different styles and colours, with the function of the helmet to provide clear visibility while blasting, together with a supply of clean fresh air. The helmet must be checked for any damages prior to and after any blasting activities.

Never leave the blast helmet inside the chamber or work area as the lining inside the helmet may become contaminated with abrasive such as lead-based contaminants which could be toxic to the operative. The helmet should have a cape attached to prevent any abrasive or dust from entering the helmet. The cape is made from a tough resistant material such as leather or thick fabric which in turn will protect the blaster's upper body.

Hand Protection

The abrasive blast cleaning operator should wear gauntlets or heavy-duty industrial gloves. The materials should be long lasting, tough and generally comfortable. The abrasive blast cleaning operator assistant should wear gloves and not gauntlets due to limited finger movement. The gloves should be checked daily for wear, tear and damage.

Although the gauntlets are manufactured for heavy duty use, they will not stand a full abrasive blast clean in full flow for more than a few seconds. The harder and sharper the abrasives, such as chilled iron grit etc will rip across the gauntlets and through to the skin in a very short period of time. Always check for wear or tear of the gauntlets before every use.

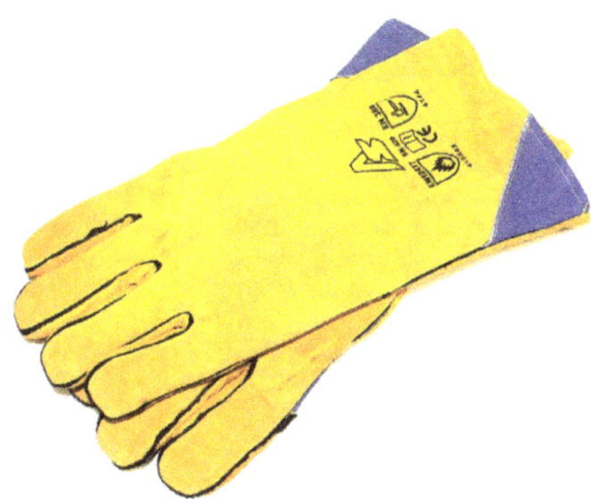

Blasting Gauntlets

Air Fed Blast Helmet

Air fed blast helmets are designed to cover the entire head and neck area and supply the blaster with clean, dry air at low pressure through a filtered air supply. They protect the wearer from dust particles including old paint coatings that may prove harmful to respiratory organs, eyes, ears and exposed skin.

The blast helmet provides the most complete means of protection because it offers ear, eye and skin protection. The continuous supply of dry fresh air prevents the helmet from misting up or fogging. The visor or mica sheets which are placed on the front of the blast helmet can be replaced when visibility becomes distorted or obscured.

Blasting Helmet with Cape

Lighting

Certain brands of blast cleaning helmets have a fixed light on top of the helmet which assists the blasting operator whilst carrying out his blasting duties; however, these are not as powerful or popular as the "hand held spot bulb". The main advantage of the helmet light is that the blast cleaning operator's hands are free, which is good from a safety standpoint; however, the hand held spot offers far more lumens and offers more visibility.

Halogen Light

Water and Oil Separator

It is essential that surface cleanliness not be impaired during the blast cleaning process. This can arise from oil vapour or droplets carried over in the air from the compressor from moisture entrapped in the high pressure air, or from residual dust arising during blast cleaning. Precautions should be taken to ensure the compressed air supplies are oil and moisture free.

The installation of suitable oil and moisture traps, together with aftercoolers and filters in the air lines, is required and these must be given adequate maintenance. Most oil and moisture traps are operated with drain plugs in a part open position, allowing accumulated moisture to disperse. Moist air can also cause abrasive to clog the abrasive lines or blast pot and may cause rusting of the blast cleaned surface.

After-Cooler

The most prevalent contaminant in compressed air is water. Water in the form of vapour, enters the air system at the compressor intake, and is concentrated to the saturation point by compression. As cooling occurs downstream of the compressor, the moisture in this saturated air condenses into harmful liquid water. Still another contaminant, oil is injected into air systems by lubricated compressors.

After-Cooler

Many litres of oil can enter an air system over the course of time in this way. Dirt takes many forms in air systems since it enters from several different sources. Small particles of dust not removed by intake filters are again concentrated by compression with the result that pipe scale forms over time.

An after-cooler on a blasting and painting project will remove all oil and water contaminants and ensure dry and clean air. The hot and moisture-laden compressed air is effectively treated to reduce and eliminate the moisture load on downstream equipment. The units are available as electrically or pneumatically operated models.

Fresh Air Filter

The air filter is an essential piece of equipment that is necessary to provide clean fresh air to the blast cleaning operator while executing his blasting duties. Without the air purifier the operator would be unable to have access to a fresh air supply and may breathe in contaminated air. A three-quarter airline runs from the compressor to the air filter unit. The filter will reduce pollutants such as water vapour and oil mist that are conveyed from the compressor, which would otherwise contaminate and pollute the operator.

A second line runs from the filter unit to the blaster's helmet, often called a "helmet lead", which is a one-quarter airline that provides fresh air to the operator, supplying him with cool fresh air.

Fresh Air Filter

Breathing air should not contain impurities in excess of the below:

- Carbon Dioxide 500 ppm (900 mg/m^3)
- Oil Mist (0.5 mg/m^3)
- Carbon Monoxide 5 ppm (5.6 mg/m^3)
- Odour and Cleanliness – the air must be free from odour and particles of abrasive or any other toxic ingredients.

Air Receiver Tank (Pig)

An air receiver tank is an integral and important part of any compressed air system. Typically a receiver tank is sized at 6–10 times the flow rate of the system, therefore is critical on blast cleaning and painting projects. If a compressor has a rating of 25 cubic feet minimum at 100 psi the receiver tank should be 150 cubic feet minimum. In a compressed air system, a receiver tank provides the following benefits:

- The receiver tank acts as a reservoir of compressed air for peak demands
- The receiver tank will help remove water from the system by allowing the air a chance to cool
- The receiver tank minimizes pulsation in the system caused by a reciprocating compressor or a cyclic process downstream.

Much like a water reservoir provides water during times of drought and stores water during the wet times, an air receiver tank compensates for peak demand and helps balance the supply of the compressor with the demand of the system. Receiver tanks are required by law to have a pressure relief valve and a pressure gauge. The relief valve should be set to 10% higher than the working pressure of the system. It is also important to install either a manual or automatic drain on the receiver tank to remove water from the system.

Air Receiver Tank

Dead Man's Handle

The dead man's handle must be held in the closed position to pressurize and therefore activate the blast pot. When the valve is released, the blast pot is shut down thus safeguarding the blast operator. If the blaster becomes ill, faints or trips and drops the blast hose, he or she is not in danger of being hit by flying abrasive or the blast hose whipping around in the work area.

The dead man's handle could be considered to be cost-effective because it can eliminate the need for a potman. Remote control valves may be operated by compressed air or electricity. Compressed air operated valves do not require any other power source than the compressor, but may be slow to react, particularly when the blast hoses and nozzles are a significant distance from the blast pot.

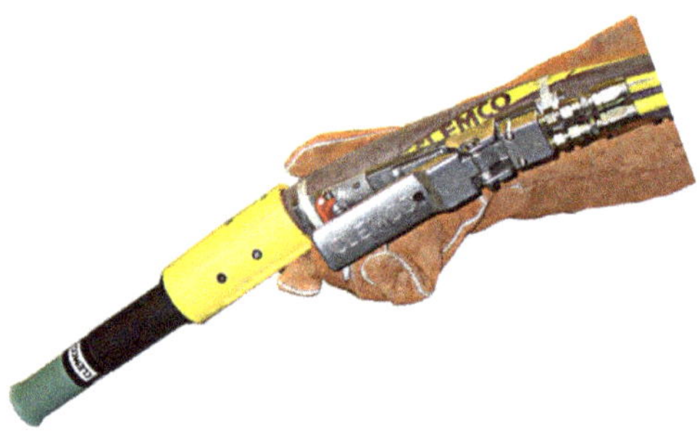

Dead Man's Handle

Electrically operated valves operate instantaneously but require a power source and are not intrinsically safe so they cannot be used in a hazardous environment. Filtered and regulated air supplied respirators are required for all dry abrasive blast cleaning. Not only the blast cleaning operative but all personnel in the contaminated area must wear approved breathing apparatus. In addition, all equipment must be grounded to prevent electric shock.

Blasting Sieve

The sieve is placed on top of the blast pot to entrap contaminants like old paint coatings, rust and scale and to keep them out of the blast pot when the abrasive is recycled. The abrasive is filtered through the sieve and then placed into the blast pot for re-use. The project specification will give instructions as to whether the abrasive can be recycled or not, and if so how many times it should be reused. The use of the sieve is normally associated with the use of metallic abrasive like steel shot and grit; however, other abrasives are sometimes used.

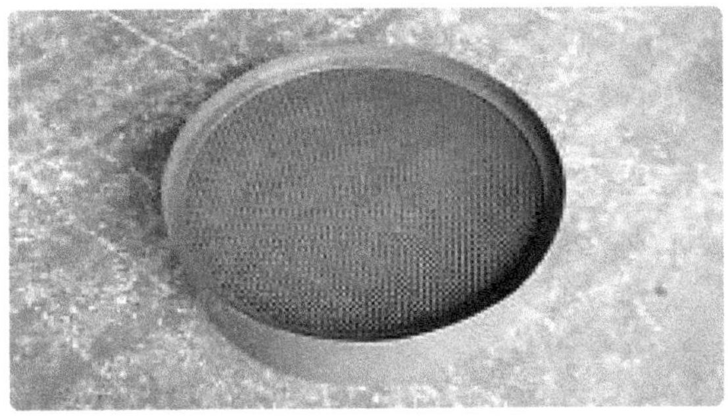

Blasting Sieve

The blast cleaned surfaces should be compared to the appropriate photographs according to the standard of ISO 8501-1. The term 'foreign matter' refers to soluble salts and welding residues. These contaminants cannot be entirely removed using dry abrasive blast cleaning. A surface containing mill scale is unsuitable for painting and must be removed by blast cleaning.

Blotter Test (ASTM 4285)

The blotter test is used to check visually for any trace of oil or water in compressed air to be used in abrasive blast cleaning, or in air used for conventional air spray application. Compressed air should be allowed to reach operational conditions.

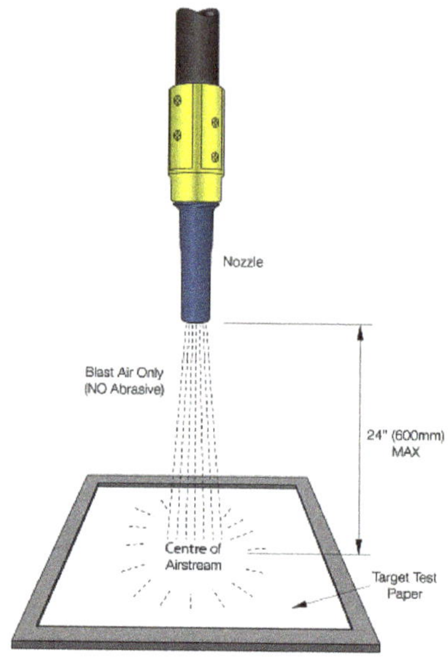

Blotter Test ASTM 4285

An absorbent collector such as a blotter paper or white cloth, or non-absorbent collector such as rigid transparent plastic should be used, and the test should be conducted without any abrasives, and the collector positioned in the centre of the airstream, 61cm (24 inches) from air discharge point, and after the in-line oil and water separators, for a period of one minute. According to ASTM D 4285 any indication of oil discolouration on the collector shall be cause for rejection.

Wet Abrasive Blast Cleaning

Wet abrasive blast cleaning is becoming more popular in the world of surface treatment due to fact that very little dust is produced during blast cleaning activities and salts can be removed. This method of cleaning was invented by Norman Ashworth, who was a pioneer for wet abrasive blasting in the 1950s.

An alternative to dry abrasive blasting was needed due to a condition called silicosis, as sand containing silica was popular in the 1950s as an abrasive media. Silicosis is a respiratory disease that can have an irreversible effect on the lungs.

Silica was banned from use in 1950 due to this condition, in the UK. The process of blast cleaning using the pressure type equipment for wet abrasive blast cleaning uses the same "set-up" as dry abrasive blasting; however, a stream of high pressure water is added to the abrasive mix at the nozzle so that dust particles are eliminated.

Wet Blast Cleaning

Some units spray water onto the abrasive as it leaves the nozzle or have a shroud on the end of the blasting nozzle to wet the abrasive. Any water soluble contaminants that are present on the substrate at the time of wet blasting are virtually eliminated. Special test equipment is available to assess such soluble contaminants [see later]. This method of blast cleaning is common on projects where dust and fines can damage or harm sensitive equipment.

Advantages of Wet Blasting

- A Surface Profile is produced
- Good for removing Lead Based Coatings
- Removes Soluble Salts
- No Dust is created
- No Sparks created.

Due to an increase in wet blasting, paint manufacturers have developed coatings which can be applied on damp and oxidized surfaces. Many a discussion has been raised regarding the degree of flash rusting acceptable for the application of coatings. Standards are available for the level of contamination remaining on the substrate prior to the application of coatings, for example the SSPC VIS 5/NACE VIS 9/ISO 8501-1.

Wet abrasive blasting can produce a high quality finish, although this is hard to make out at first as the surface is normally gingered and wet. The process is ideal for removing chlorides and will even create a surface profile. The abrasive and water help feather the edge of a repair, which eliminates the need to abrade an edge with sand paper or mechanical means.

Fresh water must be used in the wet blasting process because if seawater was used, chlorides would be forced into the surface being treated, and accelerate the corrosion process if not removed. Certain specifications stipulate that the surface is wet blast cleaned, allowed to dry, then dry blasted to a recognised standard. This method of cleaning does not come cheap.

Reminder:

- Dry abrasive blasting will not remove chlorides
- Freshwater washing will not remove soluble salts as effectively as wet blasting
- Only freshwater should be used to remove chlorides
- Nace/SSPC/ISO standards are available for wet blast cleaning.

Hand Wire Brushing

Wire brushing is one of the oldest methods of preparing a surface prior to application of coatings. The design of the wire brush has not changed much over the years – the body of the brush is made of wood with the bristles being made of ferrous or non-ferrous material.

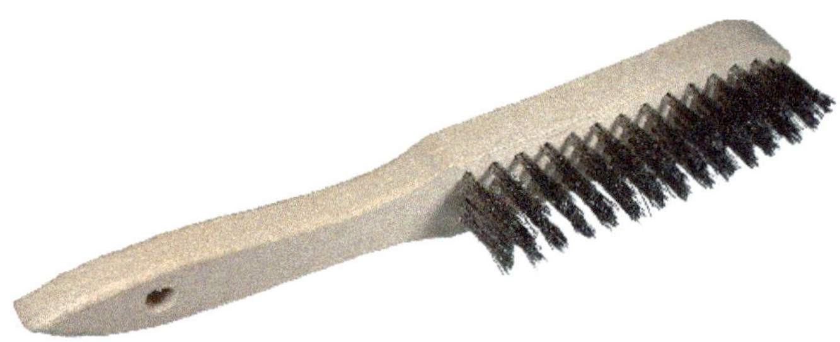

Hand Wire Brush

Copper and brass bristles are commonly used offshore in the oil and gas industry as they will not create a spark hazard. The wire brush is very simple to use and does not require an air or power source; however, it is labour intensive and a very slow process. The results of hand tool cleaning are inadequate due to the amount of contamination remaining on the surface after preparation.

Hand wire brushing will not remove intact mill scale or heavy layers of rust. Mill scale must be removed by abrasive blast cleaning, with the layers of rust being removed by needle gunning, chipping or abrasive blast cleaning. Wire brushing is known as a low level grade of surface preparation with only loose contaminants being removed. Soluble salts and grease are also not removed by hand wire brushing.

Failure to remove surface contaminants like oil, grease and soluble salts will result in loss of adhesion and premature coating failure, so it is important that these are removed prior to any surface preparation commencing. A standard commonly used to control the hand tool cleaning process is the ISO 8501-1 St 2 which is comparable to SSPC-SP 2. Visual standards such as SSPC-Vis 3, ISO 8501-1 or any other visual standard agreed upon by the contracting parties may be used to specify the surface finish required, or to verify the condition achieved.

ISO 8504-3: 2018

Descriptions of the Hand and Power Tool Cleaning grades are given below:

B St 2

B St 3

▪ St 2 - Thorough Hand and Power Tool Cleaning

When viewed without magnification the surface shall be free from visible oil, grease and dirt, and from poorly adhering mill scale, rust, paint coatings and foreign matter.

- **St 3 - Very Thorough Hand and Power Tool Cleaning**

As for St 2 but the surface shall be treated much more thoroughly to give a metallic sheen arising from the metallic substrate.

If the surface is deeply pitted, then the corrosion products lying in the bottom of the pits cannot be removed by the wire brush. Mill scale and firmly adhered paint coatings normally remain on the surface after wire brushing. Usually hand tool cleaning is only used on small areas or where power tool cleaning or blast cleaning is inaccessible.

When we mention hand tools it also refers to scrapers, chisels, chipping hammers and abrasive paper. Special formulated coatings are available for application to power tool cleaned or hand prepared surfaces. For example, epoxy mastics and surface tolerant coatings containing aluminium pigmentation perform very well on such surfaces; however, the cleaner the surface the longer the life expectancy of any coating system.

Advantages of Hand Wire Brushing

- No air or power is needed for wire brushing.
- Simple to use and operate.
- No training needed.
- Wire brushes are inexpensive.

Disadvantages of Hand Wire Brushing

- Wire brushing will not remove mill scale or heavy rust.
- Only provides short-term surface protection.
- Wire brushing will not remove grease or soluble contaminants.
- Slow method of preparation and labour intensive.
- Wire brushing will not feather an edge of a repair or existing coating.

Power Wire Brushing

Power wire brushing is another method of preparing a surface, driven by electrical or air powered means. The process of power wire brush cleaning has been around for many years and is much more effective than hand wire brushing; however, mill scale and firmly adhered scale is not removed. This method may be used to remove weld flux, weld spatter and laminations, and to smooth rough welds and round outgouges before abrasive blast cleaning.

Power Wire Brushing

Power wire brushing is widely used in maintenance painting programmes and is ideal for spot repairing areas of corrosion where blast cleaning is impractical; however, it is still labour intensive and slow compared to blast cleaning. When considering hand tool cleaning and power tool cleaning, it must be remembered that more contamination remains on the surface than with any other method of surface preparation.

The rotary wire brush will only remove loose mill scale, loose rust and loose paint coatings, it is not intended to remove all mill scale, rust and paint coatings present. Mill scale, rust and paint coatings are considered adherent if they cannot be removed with a dull putty knife. The standards most commonly used to govern the power tool cleaning process are ISO 8501–1 St 3 and SSPC–SP3.

One of the main disadvantages of power tool cleaning is that it tends to spread surface contaminants into, and around, the substrate being treated, for example oil or grease or cutting compounds. It is essential that these contaminants are removed by solvent cleaning, for example to SSPC–SP1, using lint free rags prior to any surface preparation activities.

When preparing a substrate with a rotary wire brush, the surface will burnish or polish if the cup of the brush is held on one area for too long. This will inevitably lead to loss of adhesion due to a smooth surface being obtained. Any surface profile that was once there will be reduced or even removed due to the burnishing effect of the rotary wire brush.

Needle Gun

Old paint coatings, rust and scale can be removed by pneumatic means such as a needle gun. The needle gun works by a group of needles impacting the surface around 5,000 times per minute which sharply breaks up any thick layers of coating or rust scale, which may be present on the surface. It is a simple, but effective piece of equipment using compressed air to force the piston backwards and forwards and causing the needles to break up rust scale and heavily corroded areas.

Best results are obtained by holding the needle gun at an angle of 45° as irregular surfaces can be cleaned back to bare steel in seconds. One of the main disadvantages with needle gunning is the needles pushing impurities back into the substrate, contaminating the object being treated. Another disadvantage is that needle gunning can produce an irregular surface profile when held for too long in one location.

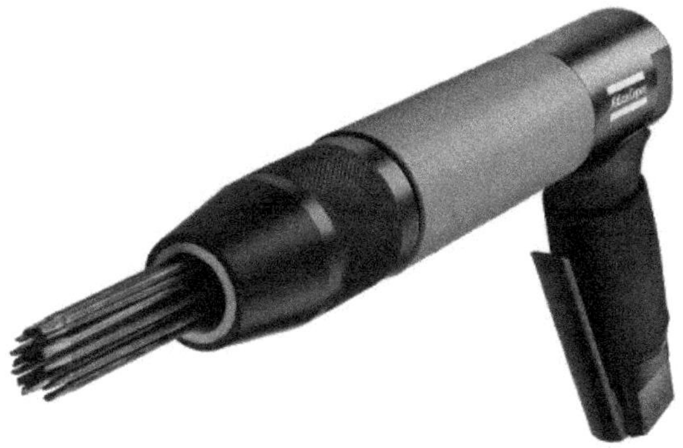

Needle Gun

Rogue peaks are peaks which stand out above the required amplitude and can lead to spot rusting if a coating is applied too thin. An issue encountered when using pneumatic vibrating tools for prolonged periods of time, especially the needle gun, is "white finger". This is an irreversible neurological injury that can lead to numbness and arthritis.

Hand and Arm Vibration (HAVS) is the medical term for symptoms caused by vibration damages that may occur in the fingers, hands and arms when working with vibrating tools. Vibration injuries are divided into three groups, neurological, vascular and musculoskeletal. Symptoms include tingling or numbness of the hands and fingers, which includes the fingers going white - hence the term "white finger".

Carpal tunnel syndrome is another common neurological vibration injury and if detected and treated early enough the patient can make a full recovery; however, fully developed neurological vibration damage is irreversible. Trigger times must be strictly adhered to when operating a needle gun and all times must be logged to prevent such injuries. Tool timers are fixed onto the body of the gun so the operator will know when his rest time, or change over time, is up, to avoid exposure to HAVS.

Scaling Hammer

The scaling hammer is the most widely used type of impact tool. A chisel is inserted into a power tool and the impact from the air operated or electrically operated piston is transmitted to the chisel, and in turn to the surface to be cleaned. Chisels may be of different shapes and constructed from different materials.

Scaling Hammer

Scaling or chipping hammers are generally a slow, costly method of cleaning the surface, but in some cases (such as when considerable rust scale or thick paint layers must be removed), they prove to be economical. Great care must be exercised in using these tools because of the tendency to cut into the surface excessively, removing sound metal and leaving sharp burrs where paint can fail prematurely.

These tools may be used to remove tight mill scale and surface rusting, but in general this is not the most practical or economical method. There is a high possibility of gouging the metal, which must be then smoothed to do a thorough job. The tools must be kept sharp, or they may drive rust and scale into the surface.

Grinders and Sanders

Grinders and sanders are often used to prepare surfaces for painting. Hand-held power sources used for this purpose may be the same as those used for power wire brushing, with suitable sanding discs or grinding wheels substituted for the wire brushes. The disc abrasive grit size used with this type of cleaning equipment must be selected properly. Too coarse an abrasive will create a deep anchor that may be unsuitable for good painting performance.

Dual Action Sanding Machine

An abrasive that is too fine may cause early clogging of the grinding wheel or sanding disc, thus making the process ineffective. Grinding is well suited for removing weld spatter, smoothing weld seams, or rounding off sharp edges or corners. Grinders are frequently used in repairing minor fabrication defects. The anchor profile produced can be extremely good with complete removal of rust and mill scale; however, such methods of cleaning are very expensive for large areas.

Hand Scraper

The hand scraper is another of the oldest methods of surface cleaning. The scraper is a tool the painter should have with him at all times on each project, and is useful in many situations from removing runs, drips and inclusions, to opening tins of paint.

Hand Scraper

Scrapers come in various forms, each offering a specific advantage. Blade widths vary from one inch to six inches and are designed for achieving productivity on many types of surfaces. The scraper handle also comes in a variety of shapes and sizes, with large handles allowing the application of more pressure, and small handles allowing tight spots to be reached.

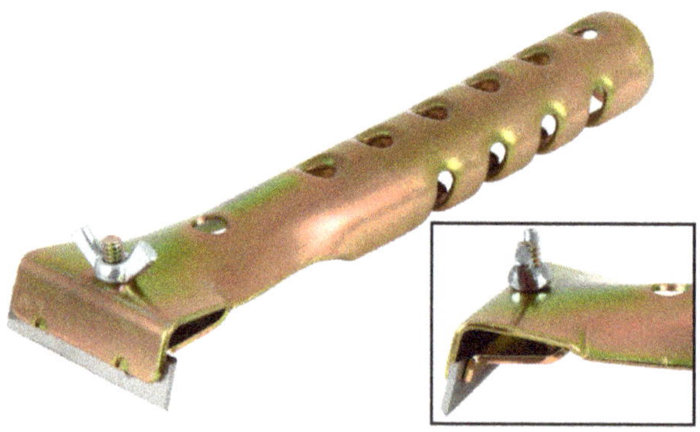

Tungsten carbide scraper

Tungsten carbide scrapers are a favourite of the industrial painter or more commonly known as "magic scrapers" as these blades provide a sharp straight edge and remain sharp for long periods of time. However, they are not indestructible and are susceptible to damage if misused or abused.

Chipping Hammer

The chipping hammer has again been used in the protective coating industry for many years, and is ideal for removing heavy layers of rust scale usually prior to blast cleaning; another industry term used is "flogging". Chipping hammers also come in electrical or pneumatic versions; however, here the most common variant, the hand held chipping hammer, will be described.

Chipping Hammer

The chipping hammer heads come in a variety of shapes and sizes; however, the double wedged shaped head is a favourite for removing paint coatings and rust scale, but it must be remembered that this operation is very labour intensive. Heavy square faced hammers are sometimes used for large, flat rust covered areas such as ships' decks or tank bottoms; however, the decks can be penetrated if heavily corroded or pitted, and also sparks are produced during the chipping process.

A rough surface may be produced with the chipping hammer as the substrate is "gouged" with the force of the hammer striking the surface, so care must be taken that the surface produced is suitable for the application of protective coatings, as premature coating breakdown could occur if "sharp edges" protrude through the paint film.

Personal protective equipment (PPE) is required when using the chipping hammer as debris such as old paint coatings, rust, scale can become airborne when being removed. PPE needed consists of goggles to protect the eyes, a dust mask to prevent the inhalation of dust and paint chippings, and ear protection, as the thudding and pounding of the square faced hammer can have an adverse effect on the hearing, especially in confined spaces (eg tanks, vessels, hoppers).

Sand Paper

Hand abrading with the use of sandpaper is sometimes used in the coating industry and is mainly used for abrading previously painted surfaces to provide a "key" to assist with coating adhesion.

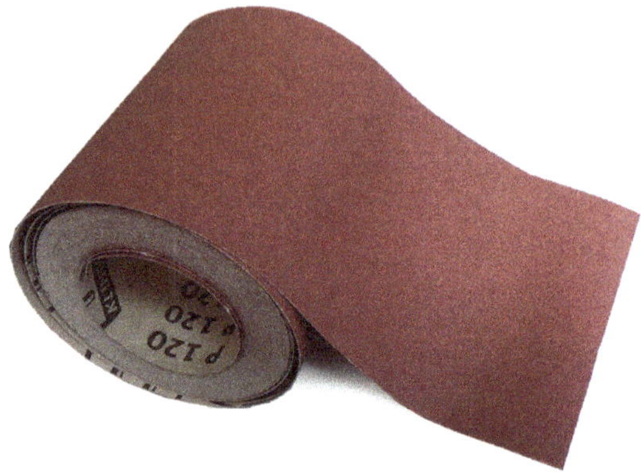

Despite the use of the name "sand paper" sand is not used in the manufacture of sand paper anymore as it has been replaced with abrasives such as aluminium oxide and silicon carbide. Sand paper is produced in a range of grit sizes with the grit size of the paper relating to the particle size, such as 20 or 40 indicating a coarse grit, while a large number such as 1500 indicates a fine grit.

Sand papers are available in a variety of grit sizes such as extra coarse (P12 – P36), Coarse (P40 – P50), Medium (P60 – P80), Fine (P100 – P120), Very Fine (P150- P220), Extra Fine (P400 – P600), Super Fine (P800 – P1200) and Ultra Fine (P1500 – P2500).

Several standards have been established for grit sizes. These standards establish not only the average grit size but also the allowable variation from the average. The two most common are the United States (CAMI) Coated Abrasive Manufacturers Institute and the European (FEPA) Federation of European Producers of Abrasives

Other standards include the Japanese Industrial Standards Committee (JIS).

Types of Abrasives used

- **Glass** – Very rarely used
- **Flint** – Very rarely used
- **Garnet** – Often used as a recyclable abrasive to prepare metal surfaces
- **Emery** – Used to polish metals
- **Aluminium Oxide** – The most commonly used abrasive
- **Silicon Carbide** – Common in wet applications.

Blasting Abrasives

In the early 1900s it was assumed that sharp edged grains of abrasive provided the best performance in cleaning and roughening a surface, but this was later demonstrated not to be correct. Mineral silica sand was used as a type of mineral abrasive, but it tends to break up quickly, creating large quantities of dust, exposing the operator to the potential development of silicosis, a debilitating lung disease, and nowadays silica is not allowed to be used.

Mineral abrasives

Mineral abrasives consist of garnet, olivine and quartz sand. Garnet, which is more expensive than silica sand, but if used correctly will offer equivalent production rates while producing less dust and no safety hazards from ingesting the dust.

Organic Abrasives

Organic abrasives typically consist of walnut shells, plastics and fruit kernels and are used to avoid damaging the underlying material such as brickwork or stone whilst removing graffiti.

Synthetic Mineral

Synthetic mineral abrasives used for surface preparation include coal slag, iron furnace slag and nickel slag. These abrasives scour and cut into the surface being treated. The scouring action is due to the fracturing of the non-metallic particle as it hits the surface, rather than having a direct impact and falling away as with steel shot.

Metallic Abrasives

Metallic abrasives comprise steel grit and shot, stainless steel, cut wire, cast iron. Steel grit is formed by crushing and cracking steel shot. Its blasting action is because of the sharp edges and is much more of a cutting action than either sand or steel shot alone, and this is especially true when the grit is new. The action of steel shot is one of impact alone. Results are similar to striking the surface with a ball peen hammer as the shot will peen and work-harden the surface.

Shot is ideal for the removal of mill scale as the energy of the heavy metal particle hitting the surface effectively cracks and disbonds the brittle rust and mill scale from steel surface. It is not, however, as efficient in removing surface residues such as mill scale binder and heavy surface rust, since these may be pounded into the surface by the peening action of the shot. Abrasive manufacturers typically relate their products to type and size, ie S 330 which is Shot (s) at 33 thousandths of an inch =.85mm. or S 220 which is shot at 22 thousands of an inch =.56mm.

Metallic Abrasives comprise the following

- **Chilled Iron Grit** – Provides a sharp angular cut surface
- **Steel Shot** – Provides a compacted and peened surface
- **Stainless Steel** – This abrasive provides excellent density and hardness
- **Cut Wire** – Provides a sharp and deep angular surface
- **Cast Iron** – Provides a sharp angular structure with excellent cleaning properties.

Chilled Iron Grit

Steel grit provides sharp and angular cutting edges and is the preferred abrasive when used in blasting chambers and centrifugal blasting systems as it can be recycled a number of times. On impact these abrasives break up to create sharp angular cutting edges, forming sharp peaks and valleys, which is ideal and well suited for coatings that require a high surface profile such as glass flake epoxies and thermal sprayed coatings.

Chilled Iron Grit

There is also some peening action by the grit particles which hit with the blunt side. As the grit is used and reused it becomes more rounded and blunt, which increases the peening action. The cutting action of the grit opens new steel surfaces and somewhat increases the reactive sites for coating attachment. This increases the adhesion potential of the coatings applied over the grit blasted surface.

A grit and shot mix is sometimes called for on certain projects. This is called a "working mix", which provides an effective and efficient blast cleaned surface. The correct working mix is important as if you use too much shot a low surface profile is created and too much grit wears out the equipment. Metallic abrasives are seldomly used on site due to the fact that reclaiming the abrasive is difficult and if contaminated with moisture the abrasive will oxidize and corrode.

Advantages of Chilled Iron Grit

- Rapid cleaning rate
- Hard abrasive – Average hardness is 57 HRC
- Low dust levels
- Abrasive breaks down slowly
- Can be re-cycled

Steel Shot

Blasting abrasives rounded in shape are known as "shot". Their common use during the development period of blast cleaning techniques leads to the common misuse of the term "shot blasting". Shot used for blast cleaning has less cutting effect and hence will prolong the life of the blast cleaning equipment especially wheel abrators. The disadvantage is the less cutting effect gives a more rounded, less roughened surface, often below the requirements of today's high performance coatings.

Steel Shot

Steel shot is still widely used in machines such as wheel abrators used to prepare steel prior to fabrication. It should be noted that any prefabrication primers used must be carefully chosen and must be compatible with the surface produced by shot blast cleaning and with the subsequent coatings if they are to remain on the steel surface and become part of the coating system. In addition to blast cleaning, shot blasting may usefully work harden a metal surface by peening, a process which can reduce the incidence of stress corrosion cracking.

Advantages of Steel Shot

- Peens and work hardens the surface
- Shatters mill scale
- Can be re-cycled
- Hard abrasive – Average hardness is 65 HRC.

Stainless Steel Abrasive

Stainless steel abrasives are very hard wearing abrasives, but due to the initial cost of the abrasive they are only used in specialised industries. Blasting with Stainless steel abrasives will not promote ferric contamination or rust on the blast cleaned substrate. This makes stainless steel an ideal abrasive choice in applications where ferrous contamination is a concern.

Stainless Steel Abrasive

Stainless steel's high bulk density and hardness results in more energy transferred to the work surface than mineral or slag abrasives, easily removing mill scale and corrosion products. The low friability of stainless steel abrasive allows it to be recycled hundreds of times; unlike mineral or slag abrasives, stainless steel will not completely shatter upon impact thus keeping its relative size.

Advantages of Stainless Steel Abrasives

- Aggressively cuts away leaving a uniform surface profile
- Low dust levels
- Will not promote ferric contamination or rust on the surface
- Easily removes mill scale.

Cut Wire Abrasive

Cut wire abrasives are carbon steel pellets chopped into small sections. The cut wire is very aggressive and therefore very effective on heavily corroded steel work or steel containing mill scale. Using chopped wire alone in a centrifugal blast enclosure will wear out the workings of the machine very quickly so mixtures of shot and grit are advisable. With the chopped wire abrasive striking the substrate aggressively a high surface profile will be produced. Due to the cost of the abrasive and difficultly in reclamation, these are limited to use in centrifugal blast cleaning and enclosed chamber work.

Cut Wire Abrasive

Cast Iron Abrasive

Cast iron abrasives are produced by grinding iron shot of a larger diameter and are available with high or low carbon content. Due to their sharp cutting edges cast iron abrasives are ideal for cleaning aggressive surfaces. When it is considered that these sharp metal particles are spinning and twisting as they are projected at the surface, it is easy to visualise sharp hackles and pinnacles of metal being created during the grit blasting process. Normally associated with use inside of a blast cleaning chamber so they can be re-cycled into the hopper then re-used, making the process effective and efficient.

Cast Iron Abrasive

Synthetic Abrasives

Synthetic abrasives consist of the following:

- **Copper Slag** – Provides high cleaning speed
- **Glass** – Re-cyclable up to 30 times
- **Dry Ice** – No dust emissions
- **Sponge** – Good for delicate machinery
- **Bicarbonate of Soda** --Relatively inexpensive
- **Aluminium Oxide** – Durable hard abrasive.

Copper Slag (Iron Silicate)

Copper slag abrasive, or iron silicate, is a by-product of copper extraction by smelting. During smelting, impurities become slag which float on the molten metal. It is commercially called copper slag since it is derived out of copper concentrates, slag that is then quenched in water produces angular granules which are disposed of as waste. Copper slag is popular with blasting contractors and is an expendable abrasive which should be used once then discarded.

Copper Slag Abrasive

The abrasive comes in a number of sizes to suit the job in hand and is ideal for site work where it is impractical to use metallic abrasives such as steel shot and grit. Generally the larger the abrasive particle size the bigger the surface profile. Copper slag abrasive is dark in colour and breaks and shatters upon impact, hence the blast cleaned surface looks darker as embedded particles are firmly lodged into the profile. The main disadvantage of this type of abrasive is the waste disposal costs, which can be classified as hazardous when removing lead based paints or other toxic coatings.

Glass Abrasive

Crushed glass is a general purpose abrasive used in a wide variety of applications. It is a manufactured abrasive made from 100% recycled glass. It is a sub-angular shape which provides a uniform surface profile and cleans by cutting through mill scale and rust while providing low particle embedment. Glass contains no heavy metals, no free silica, so is therefore good for projects where it is impractical or impossible to reclaim the spent abrasive.

Crushed Glass Abrasive

Crushed glass abrasive comes in a variety of colours and sizes and will enhance the natural lustre of the surface without any discolouration. Crushed glass is normally blasted at 90psi to achieve the most productive results. With its sub-angular shape and structure, it produces lower dust levels which in turn improves the operator's visibility, increasing production rates.

Crushed glass comes in a variety of mesh sizes ranging from extra coarse to medium to fine. Extra coarse is used to remove mill scale, rust and thick paint coatings. Medium abrasive is used for blasting aluminium and stainless steel and fine crushed glass is used to prepare surfaces in the automotive industry.

Dry Ice Abrasive

Dry ice blasting is also known as CO_2 blasting, in the solid form of carbon dioxide, which is accelerated in a pressurized airstream and directed at the substrate to clean it. As the ice hits the surface it immediately turns from a solid state into carbon dioxide vapour, chemically it sublimates, transferring minimal kinetic energy to the surface on impact, producing minimal abrasion.

Dry Ice Pellets

The sublimation process absorbs a large volume of heat from the surface producing shear stresses due to thermal shock. This improves cleaning as the top layer of dirt or contaminant is expected to transfer more heat than the underlying substrate and flake off more easily. Dry ice is a single use abrasive because it evaporates immediately after use. **Caution**: Dry ice generates very low temperatures -79 C and can cause immediate frostbite if it comes into contact with exposed skin.

Sponge Abrasive

Sponge blasting medium is a tough abrasive consisting of a sponge urethane exterior with the abrasive encapsulated in the centre of the sponge. As the sponge abrasive hits the surface it flattens, revealing the impregnated abrasive at the centre of the sponge, then as it leaves the surface the media constricts, pulling and encapsulating what would normally have become airborne contaminants. The reusable media can contain different sizes and grades of abrasives, which can be selected to prepare different substrates such as heavily corroded steel work.

Sponge abrasive has been shown to suppress up to 99% of what would be airborne dust, resulting in the need for less containment of the workplace and also reducing downtime. As sponge blasting creates very little dust, it is extremely worker friendly, and can provide a variety of profiles. Sponge abrasive comes in a variety of colours ranging from red, which is the heaviest hitter made up of steel grit, to green sponge, best for cleaning oil and grease from high value machinery.

Sponge Abrasive

Sponge abrasives come in a variety of colours, grades and sizes

Red Sponge Abrasive – Made up of Steel Grit for heavily corroded substrates

Silver Sponge Abrasive – Made up of Aluminium Oxide for aggressive substrates

Brown Sponge Abrasive – Effective for Sweep Blasting and will also leave a Surface Profile

White Sponge Abrasive – Made up of a Plastic Chip Abrasive used on exotic surfaces

Green Sponge Abrasive – is a Mild Abrasive used for difficult access areas and will leave a surface profile of up to 50µm.

Bicarbonate of Soda Abrasive

Bicarbonate of soda is a soft malleable abrasive (2.4 Mohs hardness) and is also known as baking soda with sodium bicarbonate being the chemical name. When striking the substrate, the crystals produce a unique cutting action, quickly deteriorating as it impacts the substrate. Pure sodium bicarbonate blast media effectively de-coats and cleans the substrate in one step, producing a high level of surface cleanliness not seen with most processes.

Bicarbonate of Soda Abrasive

Sodium bicarbonate blast media will not produce a surface profile, it will only reveal the existing surface profile; however, delicate substrates can be cleaned without damage. This media is not reusable, thus eliminating the issue of blasting with reused media. Recycled abrasives compound a problem by loading increasing amounts of contaminants back into the substrate.

Sodium bicarbonate blast media when applied wet is an excellent de-greasing medium that eliminates the use of solvents. Baking soda does not "dissolve" or "emulsify" grease or oils, it simply coats them, causing the grease or oil to release from the substrate.

Advantages of Bicarbonate of Soda

- Unique Cutting action
- Natural Rust Inhibitor
- Easy to dispose of
- Non-toxic
- Excellent for Degreasing when applied Wet
- Does not produce Sparks
- Use on thin or delicate substrates.

Aluminium Oxide Abrasive

Aluminium oxide is an extremely hard abrasive that is formed by a combination of aluminium and oxygen. The mineral corundum is natural aluminium oxide, and emery, ruby, and sapphire are impure crystalline varieties. It is known for its superior hardness, fast cutting action and strength, as well as its low heat retention.

Aluminium Oxide

Because aluminium oxide has the ability to cut into just about any type of substrate, it is a popular choice in a variety of blast cleaning operations. It is 50% lighter than a metallic abrasive and cuts even the hardest of metals and sintered carbide. Reclaiming is critical due to the cost of the abrasive which can be recycled many times over and widely used due to its longevity and hardness. The larger the abrasive size, the faster the aluminium oxide will cut.

Natural Abrasives

Natural abrasives consist of the following:

- **Garnet** – Fast Cutting Abrasive
- **Walnut shells** – Excellent for Graffiti Removal
- **Olivine** – Hard Silicate Mineral.

Garnet Abrasive

Comes from the Greek word "Gernet" meaning dark red. Garnet comes in many colours including brown, purple, orange, yellow, green, with red being the most common. Garnets are a group of silicate minerals that have been used for a number of years in the protective coating industry. Unlike some blasting abrasives, garnet does not contain high levels of toxic substances known as carcinogens; this makes garnet a much safer abrasive to use for the workers operating the blasting equipment and the personnel around the site.

Garnet Abrasive

It is a natural hard abrasive found in various locations of the world from India to Australia and is very environmentally friendly. Garnet is a robust abrasive which can produce good cleaning results with a high surface profile even at low pressures and is ideal for conducting surface preparation on spot repairs and field welds.

Due to its hardness (Mohs 6.5 – 7.5) and sharpness, it is ideal for feathering the edges of a coating repair, especially high build coatings as this will reduce the need for hand abrading or grinding.

Walnut Shells Abrasive

Walnut shell abrasives are a natural abrasive made entirely from crushed walnut shells. Coarse shells produce a tough, versatile abrasive while the medium and fine abrasive shells are relatively soft so there is little chance to damage adjacent substrates. When used as a blast media walnut abrasive is very durable, angular and multi-faceted, yet considered a soft abrasive. The exploding abrasive has a different cleaning action from a metallic abrasive.

The breaking up of the walnut particles into many pieces on impact tends to send the broken particles of walnut deep down into the pitted surface resulting in a contamination free surface. The scouring action removes the fine rust and corrosion from flat areas as well as pits, much more effectively than steel shot or grit which does not shatter or break up on impact.

Crushed Walnut Shells

Walnut shell blasting is an excellent replacement for sand (which is banned by most developed countries to avoid inhalation health concerns). With being a soft abrasive and lightweight, it is effective in removing a top coat while leaving the under-coat intact.

Walnut Shell Abrasive Advantages:

- Cost effectiveness as the media can be re-used
- Safe to use as it poses no health risk
- Environmentally Friendly.

Olivine Abrasive

Olivine is a naturally occurring silica-free blasting media of low specific density and has proven very popular with expendable users. With olivine being iron free and containing no free silica, it is ideal for use on stainless steel and aluminium surfaces. It is an inert and non-toxic naturally occurring mineral abrasive which has been approved in many areas as suitable for use on structures over rivers and water courses, and is a safe alternative to sand.

Olivine Abrasive

Olivine is extremely good for wet blasting operations as it does not leave any trace marks on the surface of the substrate being treated, unlike iron silicates. The cleaning action of olivine is less effective on heavily rusted areas than metallic abrasives such as steel or grit. Once the heavy scale is removed, usually by a chipping hammer or chisel, the cleaning action of olivine is superior. Indian olivine is known for having the highest (7 Mohs) hardness and lowest loss on ignition makes it an ideal media for blasting and water-cutting.

CHAPTER THREE
Surface Profile

In addition to achieving a clean surface, abrasive blast cleaning alters the substrate from a smooth surface to a uniformly textured surface. This textured surface is the result of the sharp abrasive particles striking the steel substrate at high speed and leaving small impact indentations. This texture is called surface profile or anchor pattern. Surface profile is an important factor to consider in the protective coatings industry as it is the profile that determines the mechanical adhesion of a coating.

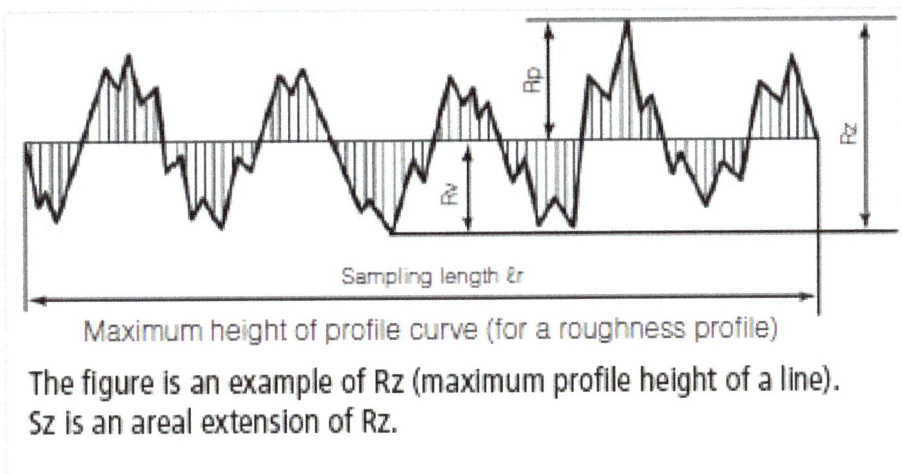

Example of Rz

When abrasive blast cleaning is used to clean a substrate, the surface area is increased, leading to an increase in paint consumption as peaks and troughs need filling with coating. If the substrate has a high surface profile it will require a thicker layer of coating to ensure that all of the peaks are fully covered because if the peaks are not fully covered "Spot Rusting" will occur, leading to coating failure. Too shallow a surface profile may result in premature coating failure due to lack of adhesion seen as peeling.

Using large particle size abrasives for blast cleaning will produce a high surface profile. The greater the surface profile the greater the coating adhesion will be. One exception to this rule appears to be inorganic zinc silicate primers, which tend to split (lose cohesion) when the anchor profile exceeds about 67 µm. This effect may be caused in part by attempts to increase the thickness to achieve coverage of the profile peaks. Inorganic zinc silicate primers are well known to be sensitive to excess thickness. The guidelines provided by the coating manufacturer on technical data sheets should be followed.

Thermal sprayed coatings and glass flake epoxies are commonly applied at high thicknesses so require a rough or high profile. Not only do these coatings require a high surface profile they also require a higher degree of surface cleanliness such as an ISO 8501 Sa 3. Mechanical adhesion is the key to a successful coating system, therefore amplitude and cleanliness are paramount.

Quality control must be strictly observed when using large abrasive particle size as rogue peaks may also be produced. These are peaks which stand out above the required amplitude due to blasting in one area for too long. If these peaks are painted over, they will lead to "pin point rusting" and eventually coating failure.

Surface Profile Measurement

Several methods have been developed to measure the peak to trough height or amplitude of the surface profile, and these include:

- Surface Comparator
- Replica Tape
- Surface Profile Needle Gauge
- Digital Surface Profile Gauge.

Surface Comparator

Surface profile can be assessed visually with the use of a surface comparator. The comparator is available in both grit (Type G) and shot (Type S) versions. Limits of profile grades are fine, medium and coarse. This method uses a square panel with a central hole surrounded by four segments with different grades of roughness. With the assistance of a 5X (and not exceed 7X) lighted magnifier, the comparator is placed against the substrate then visual and tactile comparisons are made.

Five Grades may be Recorded:

- Finer than fine grade
- Fine grade
- Medium grade
- Coarse grade
- Coarser than coarse grade.

Surface Comparator

The comparators are referred to in **ISO 8503-2. Preparation of Steel Substrates before the application of Paints and Related Products – Surface Roughness Characteristics of Blast Cleaned Substrates.**

- Remove all loose dust and debris from the test surface.
- Select the appropriate surface profile comparator and place it against an area of the surface.
- Compare in turn the test surface with the four segments of the comparator using the hand lens (magnifying glass) if necessary.
- If the hand lens is used, place it so that the test surface is viewed simultaneously with a segment of the comparator.
- Assess the profiles on the comparator that are nearest to the profile of the test surface and from these determine its grade.

Limits of Profile Grades:

- **Fine** – Profiles equal to Segment 1 and up to but excluding segment 2
- **Medium**–Profiles equal to segment 2 and up to but excluding segment 3
- **Coarse** – Profiles equal to segment 3 and up to but excluding segment 4.

Note – If visual assessment proves difficult it is possible to assess the closest grading by passing either the back of the finger nail or a wooden stylus held between the thumb and forefinger alternately over the test surface and the segments of the comparator.

- Repeat the test with the comparator adjacent to each of the areas of the test surface as required.
- Record the grades for all areas of the test surface.
- If any profile is assessed as below the lower limit for the "fine" grading, repeat the grading as "finer than fine".
- If any profile is assessed as above the upper limit for the "coarse" grading, report the grading as "coarser than coarse".

Testex Tape (Replica Tape)

Testex tape or replica tape is a commonly used test conducted in the coatings industry for assessing surface profile. Three types of tape are used: coarse, extra coarse, extra coarse plus. The backing is removed from the reverse side of the tape to reveal an adhesive layer.

The tape is then placed on the blast cleaned substrate and a burnishing tool is then used to compress the mylar tape into the surface profile, the shape of the profile is permanently impressed onto the tape. The tape is then placed into the dial micrometer for measurement. The tape has a 50μm (2 mil) mylar layer on the back which is subtracted when measuring with the dial micrometer. The testex tape can then be kept as permanent record within a coating inspection report.

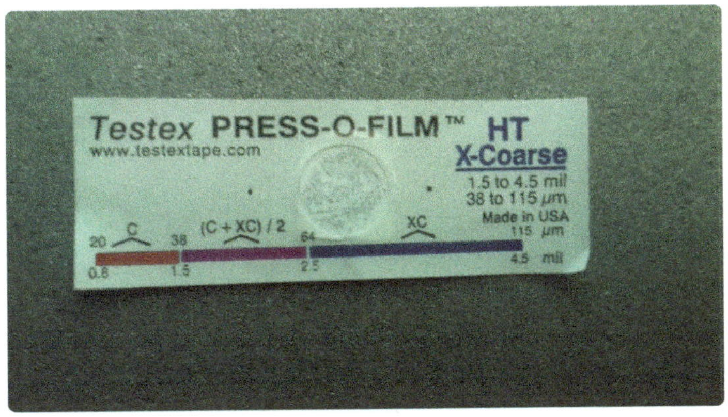

Testex Tape

Testex Tape or Replica Tape is available in certain grades:

- Coarse – 20µm to 64µm
- Extra Coarse – 38µm to 115µm
- Extra Coarse Plus – 116µm to 127µm.

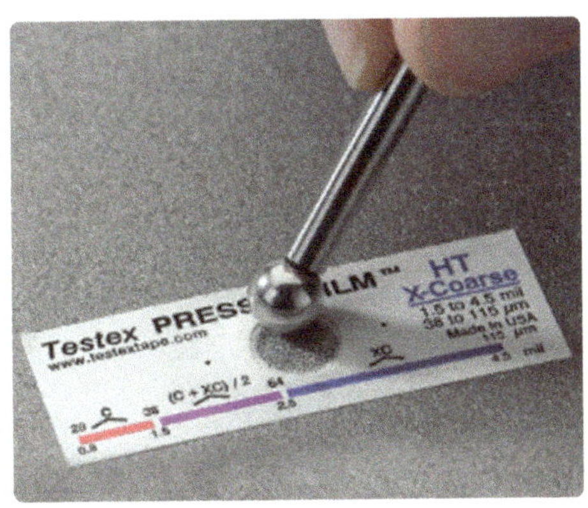

A Burnishing Tool applied to the Testex Tape

Procedure for Measuring Surface Profile

Preparation of Steel Substrates before application of Paints and Related Products – Surface Roughness Characteristics of Blast Cleaned Steel Substrates – Part 5 Replica Tape Method for the Determination of the Surface Profile: ISO 8503-5

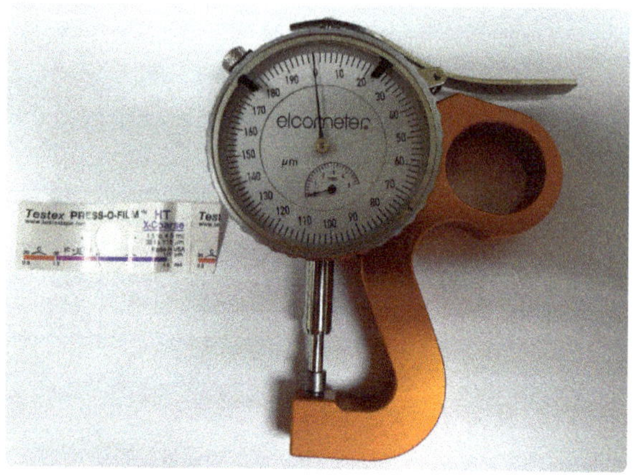

Dial Micrometer

- The micrometer is set to zero prior to any readings being taken.
- The plastic backing is removed from the film then the testex tape is placed on the clean blasted substrate and is rubbed with a circular ended tool until the surface has conformed to that of the steel indicated by a uniform dark colouration.
- The tape is then removed and measured with the dial gauge.
- The maximum profile can then be calculated by subtracting the thickness of the mylar backing 50μm (2 mil) from the dial reading.
- The replica tape method is very easy to use, especially on difficult to access areas.

When measuring the surface roughness inevitably there will be some rogue peaks on the substrate that need to be taken into consideration. These can be described as peaks that stick out above the required amplitude and can cause spot rusting if a thin coat is applied over these high peaks.

Surface Profile Needle Gauge

The surface profile is important when abrasive blast cleaning. There are various methods of testing or checking the surface profile. One of the oldest forms of checking the surface profile is with a surface profile needle gauge. This gauge measures the peak-to-valley height of a blast cleaned surface.

Surface Profile Needle Gauge

The needle gauge is placed onto the steel surface where the needle falls into the surface profile. The profile measurement can be read directly from the gauge. The needle is quite delicate and care must be taken when using it and should be cleaned and put away after use. A number of readings is necessary to achieve an average surface profile.

The ASTM D4417-B is the Standard Test Method for Field Measurement of Surface Profile

Using a Surface Profile Needle Gauge:

- Ensure that the gauge is set to zero before taking any measurements.
- Place the gauge on the surface to be measured, the spring tip is projected into the valley and the depth is indicated on the dial in microns or mils.
- To obtain a representative value of peak to valley height, a number of readings (at least 10) should be taken on a given surface area and the average obtained.
- When evaluating the results, it must be remembered that the instrument reads from the three maximum peaks under the foot.

Digital Surface Profile Gauge

Digital surface profile gauges are used to measure the profile height of a blast cleaned surface. The profile gauge records the value from the top of the peak to the bottom of the valley, measuring the roughness over a specified distance. These gauges provide more of a sophisticated approach into recording readings digitally without having to write it down.

One of the main advantages of the digital surface profile gauge is that surface profile can be measured on either a flat or curved surface.

How to use the Digital surface profile gauge:

- Ensure probe is 90° to substrate to ensure accurate readings and calibrate the gauge on a glass zero tile.
- To take one spot measurement, take either 5 or 10 readings within a 100mm (4") diameter area.
- The measurement displayed is either the average peak-to-valley height or the maximum reading of the 5 or 10 readings, depending on which measurement mode you have selected.
- To establish the average surface profile of an area, record 5 spot measurements for each 1m x 1m (3ft x 3ft) area.

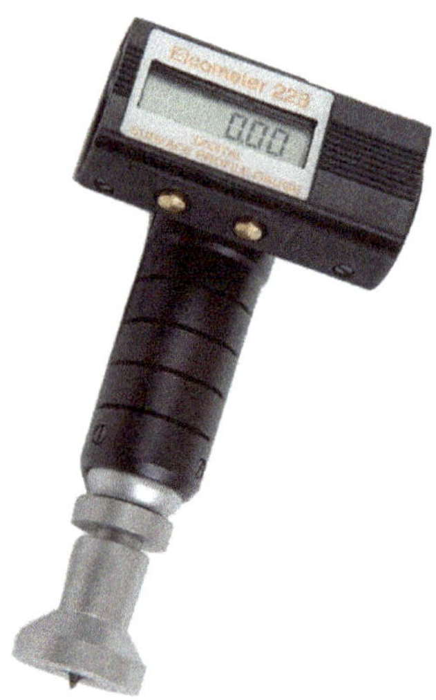

CHAPTER FOUR
Surface Preparation Standards

Visual standards are produced by several organisations including NACE, SSPC and ISO. NACE (National Association of Corrosion Engineers) and SSPC (Society for Protective Coatings) previously known as the Steel Structures Painting Council and the International Organization for Standardization (ISO).

Steel Condition

The amount of time required to achieve any particular degree of surface preparation depends on the condition of the substrate being treated. The amount of mill scale, rust, dust, dirt and paint coatings also needs to be taken into consideration.

ISO 8501-1

The standard ISO 8501–1:2007 is used for visual assessment of rust grades and surface preparation grades. It is a hardback book that includes 28 photographs of rust grades, blast cleaning grades and hand and power tool cleaning grades. The title of the book is: **Preparation of Steel Substrates before.**

The Rust Grade Book

Application of paints and related products – Visual assessment of Surface Cleanliness. Part One: Rust Grades and Preparation Grades of uncoated Steel substrates and of Steel substrates after removal of previous coatings (ISO 8501–1:2007)

ISO 8501-1: Rust Grades

Four rust grades designated A, B, C and D respectively, are specified. The rust grades are defined by written descriptions.

A – Steel surface largely covered with adhering mill scale but little, if any, rust.

B – Steel Surface which has begun to rust and from which the mill scale has begun to flake.

C – Steel Surface on which the mill scale has rusted away from which it can be scraped, but with slight pitting visible under normal vision.

D – Steel Surface on which the mill scale has rusted away and on which general pitting is visible under normal vision.

A
Steel surface largely covered with adhering mill scale but little, if any, rust

B

Steel surface which has begun to rust and from which the mill scale has begun to flake.

C

Steel surface on which the mill scale has rusted away from which it can be scraped, but with slight pitting visible under normal vision.

D

Steel surface on which the mill scale has rusted away and on which general pitting is visible under normal vision.

The Rust Grades should be assessed prior to conducting the surface preparation as the surface preparation standard must be included in the pre-surface condition.

An example of the pre-surface condition:

- Rust Grade B
- Blast Cleaning Preparation Standard – Sa3
- Standard Required – B Sa3.

ISO 8501-1:Blast Cleaning Standards

The blast cleaned surfaces should be compared to the appropriate photographs according to the standard of ISO 8501–1. The term "foreign matter" refers to water-soluble salts and welding residues. These contaminants cannot be entirely removed using dry abrasive blast cleaning.

A surface containing mill scale is unsuitable for painting and this must be removed by blast cleaning. This standard refers to the surface appearance of the steel after abrasive blast cleaning and gives descriptions with pictorial references of the grades of cleanliness.

Grades of Cleanliness are presented in the ISO 8501–1 2007 Preparation of Steel Substrates before application of Paints and Related Products – Visual Assessment of Surface Cleanliness

- **Sa 1** – Light Blast Cleaning
- **Sa 2** – Thorough Blast Cleaning
- **Sa 2.5** – Very Thorough Blast Cleaning
- **Sa 3** – Blast Cleaning to Visually Clean Steel.

The ISO 8501-1 Standard Grades of Surface Cleanliness for Abrasive Blast Cleaning are as follows:

Sa 1

Sa 2

Sa 2 1/2

Sa 3

Sa 1: Light Blast Cleaning: When viewed without magnification the surface shall be free from visible oil, grease, dust, dirt, and from poorly adhering mill scale, rust, paint coatings, and foreign matter.

Sa 2: Thorough Blast Cleaning: When viewed without magnification the surface shall be free from visible oil, grease, dust, dirt, and from most of the mill scale, rust, paint coatings, and foreign matter. Any residual contamination shall be firmly adhering.

Sa 2.5: Very Thorough Blast Cleaning: When viewed without magnification the surface shall be free from visible oil, grease, dust, dirt and from mill scale, rust, paint coatings, and foreign matter. Any remaining traces of contamination shall show only as slight stains in the form of spots and stripes.

Sa 3: Blast Cleaning to Visually Clean Steel: When viewed without magnification the surface shall be free from visible oil, grease, and dirt, and shall be free from mill scale, rust, paint coatings and foreign matter. It shall have a uniform metallic sheen.

ISO 8504-3: 2018
Hand and Power Tool Cleaning Grades

St Grades are associated with power tool cleaning and Sa grades associated with blast cleaning. There is no St grade for rust grade A.

B St 2

St 2 - Thorough Hand and Power Tool Cleaning

When viewed without magnification the surface shall be free from visible oil, grease and dirt, and from poorly adhering mill scale, rust, paint coatings and foreign matter.

B St 3

St 3 - Very Thorough Hand and Power Tool Cleaning

As for St 2 but the surface shall be treated much more thoroughly to give a metallic sheen arising from the metallic substrate.

Flame Cleaning F1

When viewed without magnification the surface shall be free from rust, mill scale, paint coatings, foreign matter. Any remaining residues should show only a discolouration of the surface. Flame cleaning is included in the ISO 8501-1 and after completion of the flame cleaning process mechanical cleaning in the form of power tool cleaning is normally encountered.

Before the process of flame cleaning any heavy layers of rust and scale should be removed by flogging or chipping. This process is normally conducted with heavy square headed hammers or pneumatic chisels. All visual assessments should be conducted in daylight or with good illumination. Check the steel condition and compare it with the photograph in the standard whilst holding the standard as close as possible to the surface being assessed.

NACE/SSPC Joint Standards

The SSPC Vis 1 is used for visual assessment just like the ISO 8501-1, as it shows the appearance of both previously unpainted and previously painted and partially rusted hot rolled carbon steel, prior to and after abrasive blast cleaning.

These joint blast cleaning standards (NACE/SSPC) will determine conformance with the blast cleaning requirements. The Vis 1 standard needs to be held as close as possible to the surface being prepared and under good daylight or illumination so an accurate description can be obtained.

The SSPC Vis 1-89 visual standard for abrasive blast cleaned steel begins by providing colour photos of four steel conditions:

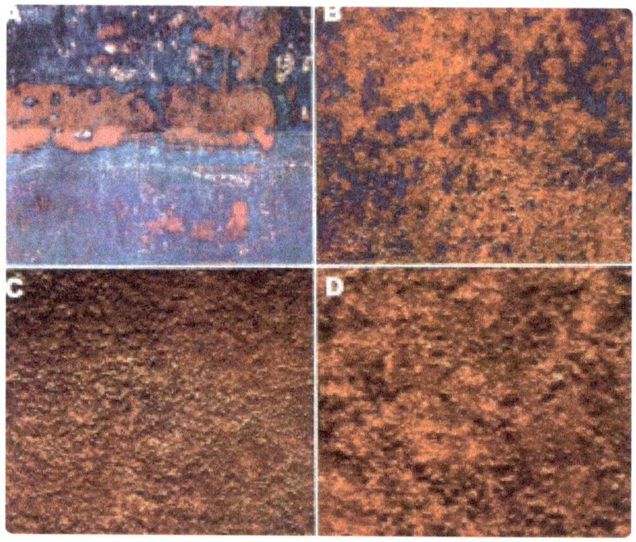

Rust Grade Condition A: Steel surface completely covered with adherent mill scale with little or no rust visible.

Rust Grade Condition B: Steel surface covered with both mill scale and rust.

Rust Grade Condition C: Steel surface completely covered with rust with little or no pitting visible.

Rust Grade Condition D: Steel surface completely covered with rust and pitting visible.

NACE/SSPC Joint Blast Cleaning Standards

NACE 1/SSPC-SP5 - White Metal Abrasive Blast

A white metal blast cleaned surface. When viewed without magnification the surface shall be free of all oil, grease, dust, dirt, mill scale, rust, coating, oxides, corrosion products, paint and other foreign matter.

NACE 2/SSPC-SP10 - Near White Abrasive Blast

The near white abrasive blast cleaned surface (NACE 2/SSPC – SP10), when viewed without magnification the surface shall be free of all oil, grease, dust, dirt, mill scale, rust, coating, oxides, corrosion products, paint and other foreign matter except for staining which shall be limited to no more than 5% of each unit area of surface (9in/6400mm), which may consist of light shadows, slight streaks or minor discolouration caused by stains or rust, stains of mill scale or stains of previously applied coatings.

NACE 3/SSPC - SP6 - Commercial Abrasive Blast

Commercial blast is defined as a surface viewed without magnification, the surface shall be free of all oil, grease, dust, dirt, mill scale, rust, coating, oxides, corrosion products, paint and other foreign matter except for staining, which shall be limited to no more than 33% of each unit area of surface (9in/6400mm) which may consist of light shadows, slight streaks or minor discolouration caused by stains or rust, stains of mill scale or stains of previously applied coatings.

NACE 8/ SSPC-SP14 - Industrial Blast Cleaning

Industrial Blast Cleaned Surface: An industrial blast cleaned surface, when viewed without magnification, shall be free of all visible oil, grease, dust, and dirt. Traces of tightly adherent mill scale, rust, and coating residues are permitted to remain on 10% of each unit area of the surface (approximately 5,800 mm^2 [9.0 in.2]) (i.e., a square 76 mm x 76 mm [3.0 in. x 3.0 in.]) if they are evenly distributed. The traces of mill scale, rust, and coating are considered to be tightly adherent if they cannot be lifted with a dull putty knife. Shadows, streaks, and discolourations caused by stains of rust, stains of mill scale, and stains of previously applied coating may be present on the remainder of the surface.

NACE 4/ SSPC-SP7 - Brush Blasting (Sweep Blasting)

Is defined as a brush blast cleaned surface when viewed without magnification, the surface shall be free of all visible oil, grease, dirt, dust, loose mill scale, loose rust and loose coating. Tightly adherent mill scale, rust and coating may remain on the surface. Mill scale, rust and coating are considered tightly adherent if they cannot be removed by lifting with a dull putty knife.

SSPC Power Tool Cleaning/Hand Tool Cleaning Standards

SSPC SP-11 - Power Tool Cleaning to Bare Metal

When viewed without magnification, the surface shall be free of all visible oil, grease, dust, dirt, mill scale, rust, paint, oxide, corrosion products and other foreign matter. Slight residues of rust and paint may be left in the lower portions of the pits if the original surface was pitted. It also requires a new profile of at least 1mil (25 Microns)

SSPC SP–2 specification is essentially the same as SSPC SP–3 except that the only power used is provided by the labourer's arms and hands.

SSPC SP-2 - Hand Tool Cleaning

All loose mill scale, loose rust, loose paint and other loose detrimental foreign material shall be removed. Mill scale, rust and paint are considered adherent if they cannot be removed by lifting with a dull putty knife.

SSPC SP-3 - Power Tool Cleaning

All loose mill scale, loose rust, loose paint and other loose detrimental foreign material shall be removed. Mill scale, rust and paint are considered adherent if they cannot be removed by lifting with a dull putty knife.

SSPC SP-1 - Solvent Cleaning

The SSPC definition for solvent cleaning (SSPC–SP1) is a method for removing all visible oil, grease, soil, drawing and cutting compounds and other soluble contaminants from steel surfaces. Emulsion or alkaline cleaners may be used in place of solvents, provided that the surface is washed with fresh water or steam to remove detrimental residues.

CHAPTER FIVE
Surface Cleanliness

During the blast cleaning process fine dust particles are created as the abrasive shatters upon impact as it hits the surface being treated. If these dust particles are not removed prior to coatings application, adhesion failure is inevitable. It is essential that all dust/debris contamination is removed by compressed air or vacuum system from the substrate being treated.

After blast cleaning activity the blast cleaning operator is responsible for cleaning down all surfaces to remove spent abrasive and light dust prior to the application of protective coatings. It may take quite a while for all of the fine airborne dust particles to settle on the surface after blast cleaning, so it is essential that the inspectors understand the importance of surface cleanliness.

The dust particles can be very fine and also very hard to see especially in areas of poor lighting and extraction so through cleaning is required. The coating will be unable to "flow" or "wet" if dust is still present in the troughs of the profile.

Assessment of Dust

If dust and grit are left on the surface prior to coating application the coating will not thoroughly "wet" the substrate. Once the coating has dried it will lose adhesion from the substrate. It is essential that applicators and inspectors fully understand the implications of inadequate cleaning.

The inspection process must follow through from visual inspection of the substrate to the cleanliness inspection. It may be some time after abrasive blast cleaning, before all the dust settles on the horizontal surfaces.

Specialised equipment such as grit recovery systems and dust removal systems are often used during and after abrasive blast cleaning. Light dust usually settles on horizontal surfaces and final cleaning is required. These dust extraction units are available in a number of sizes to suit the application, from factory units to smaller site units. It is essential that adequate dust collection is used on a continuous basis throughout the abrasive blast cleaning process.

Uncontained emissions from surface preparation operations, such as blast media, paint coatings, and rust, can be toxic and lethal if discharged and become airborne into the atmosphere. Containment is used to prevent blast emissions and debris from leaving the worksite, where they can be effectively collected for disposal.

Ideally there should be about eight air changes per hour inside of the containment. This will ensure the abrasive blast cleaning operative can see clearly and the area remains a safe area. The equipment should remain in operation during both the abrasive blast cleaning up to the final cleaning.

A light blow down with clean dry air is sometimes conducted. However, beware of blowing the dust from one area to another. Full personal protection equipment is required for this task. Vacuum cleaning for final clean is often specified and conducted especially in internal tank and vessel painting. This process ensures all dust and debris is removed prior to coating operations. Try and remove dust in a logical fashion starting at one end and finishing at another.

Vacuum cleaning can be very effective especially to horizontal flat surfaces. All the dust is collected from within the surface profile. The vacuum units can be quite powerful and are available with different heads to suit the task. Regular cleaning of the equipment is often necessary. Surface dust may be impossible to see with the naked eye, so a dust tape test kit is used to assess the amount of dust contamination on the blast cleaned substrate.

This test is in accordance with **ISO 8502-3** Preparation of Steel Substrates before application of Paints and related products – **Assessment of Dust on Steel Surfaces prepared for Painting – Pressure Sensitive Tape Method.**

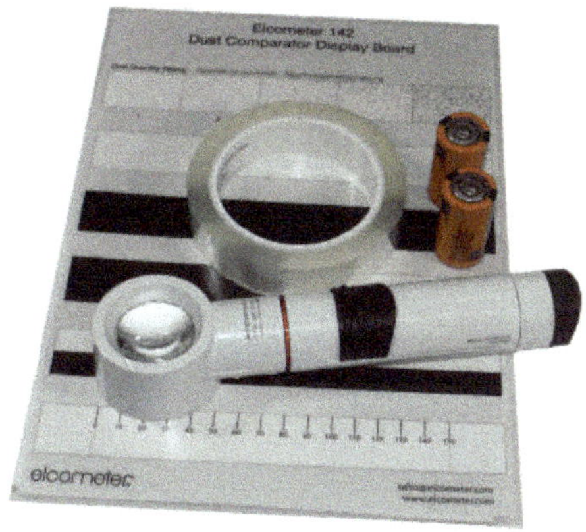

Dust Test Kit

The Procedure is as Follows Below:

- At least 150mm of sticky tape is placed on the blast cleaned substrate and rubbed three times in each direction, the tape is then placed onto the transparent strip.
- Remove the tape from the substrate.
- Place tape on the plastic strip.
- Place the plastic strip onto a black or white strip on the comparator display board.
- Assess the quantity of dust on the tape, compare the dust on the tape with the samples on the bottom of the comparator display board and record the dust rating number which is the closest match.
- Assess the size of the dust particles on the tape, view the dust using the magnifier and record the dust size class of the particles.
- Carry out at least three tests on the surface being investigated. If the dust rating number of the three tests differs by more than one, conduct at least two more additional tests and calculate the average of the results.
- **NOTE**: It is surprising the number of coating inspectors that have conducted this "assessment for dust test" placing only ONE piece of adhesive tape to the blast cleaned substrate. The ISO 8502-3 standard states "At least three tests on the surface being investigated". If the dust rating number of the three tests differs by more than one, conduct at least two more additional tests and calculate the average of the results.

Salt Contamination

Chlorides

Even after blast cleaning operations water soluble contaminants such as chlorides could still be present on the substrate especially if the surface is pitted to Rust Grade C or Rust Grade D. The abrasive is unable to reach the retained chloride sitting in the bottom of the pit and even with the use of a fine blasting medium it is unlikely to be removed, therefore testing is required to detect the presence of any remaining salt.

When water soluble salts are present on the substrate and the coating is in contact with an aqueous environment, osmotic blistering can occur as water is absorbed through the coating and transported to the location of the salts.

Coating specifications will specify the maximum amount of salts allowed, for example NORSOK M-501 specifies 20mg/cm^2; however, all client specifications are different. If the maximum level of salts is exceeded, the surface is normally washed with freshwater. Sodium chloride contamination is mainly found in a marine/coastal environment, for example shipyards and offshore structures.

The problem with steel and chlorides is the formation of ferrous chloride whenever the two materials come into contact. When a chloride ion is absorbed into rust and held onto the steel surface, the iron ion, the chloride ion and water form an iron chloride salt solution which is not only conductive and can aid in the electrochemical corrosion reaction, but is in itself a strong corrodent of the steel surface.

The ferrous chloride oxidizes upon exposure to air to form ferric chloride. Ferric chloride is a hygroscopic salt which tends to draw moisture from humidity in the air, creating a ferric chloride solution on the steel surface. In this case even though the steel work is thoroughly grit blasted, even a few molecules or ferric or ferrous chloride remaining on the surface or in the area of a pit tend to accumulate moisture from the air, causing a concentrated iron chloride solution.

Organic coatings applied over such a surface would tend to fail by the iron salt drawing water through the coating by osmosis and creating an area of poor adhesion or blistering. Wet abrasive blast cleaning or water jetting is a more effective measure of removing soluble chlorides.

Sulphates

Sulphate contamination is commonly found in industries such as refineries, petrochemical processing plants and marine environments. The marine industry can develop sulphate contamination through the breakdown of marine organisms such as barnacles and oysters. Sulphates can react directly with a steel substrate without the need for moisture. They can form a heavy iron scale, which when it has access to oxygen oxidizes very rapidly to iron oxide.

Sulphate Contamination

Iron sulphate is cathodic to steel, therefore accelerates the corrosion process. When sulphates are present on a surface they are very difficult to remove even with abrasive blast cleaning. Sulphate contamination will remain on the substrate even when blast cleaned to Sa3 white metal finish.

Procedure to Remove Sulphate Contamination is as follows:

Blast clean to Sa3 white metal finish then the surface is washed with freshwater and allowed to flash rust. Once the surface is dry it is then blast cleaned to Sa3 white metal finish. The surface is then checked for sulphate contamination using the appropriate method/kit for assessing salts. For example, the Elcometer 134 chloride, sulphate, nitrate kit, can accurately measure sulphate contamination along with other salt contamination.

Osmotic Blistering

Osmosis concerns the passage of moisture through a coating. To be more precise "osmosis" is the passage of water through a semi-permeable membrane from a solution of less concentration to one of greater concentration. All organic coatings are semi-permeable membranes and liquids can diffuse through them.

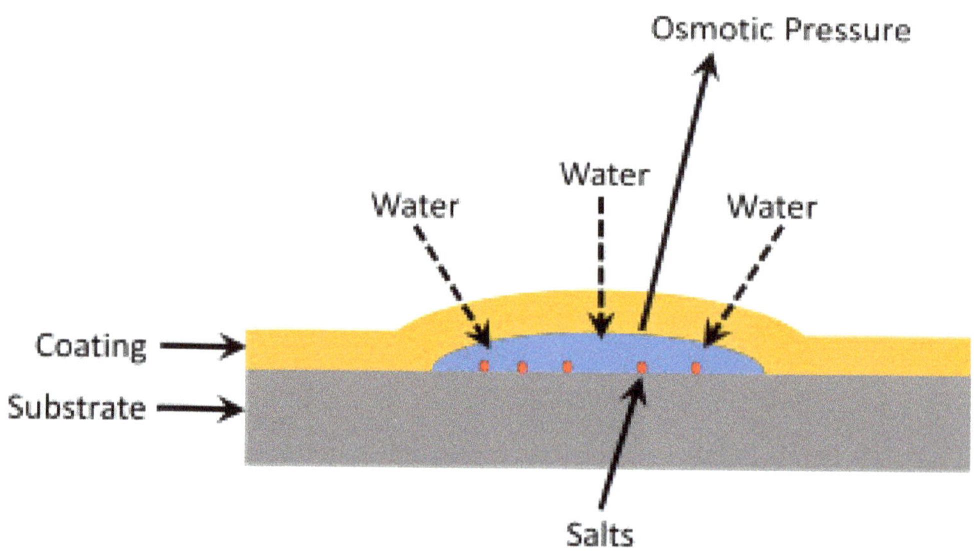

Example of Osmotic Blistering

The coatings will attain equilibrium with its environment by absorbing or releasing water depending on the prevailing relative humidity and the rate and degree to which this occurs will be influenced by factors such as the generic type of coating and the pigmentation, where laminar pigments such as MIO provide more of a barrier than normal pigments. Where water soluble salts are present at the coating/substrate interface or between coats in a multicoat system and the coating is in contact with an aqueous environment, blistering can occur as water absorbed by the coating and transported to the location of the salts.

Osmotic Blistering

If a hand print or a drop of sweat is present on a blast cleaned surface and the surface contaminant has dried it is often missed by the applicator. However, once the coating is put into service under immersion conditions or high humidity, the moisture vapour will transfer through the paint film and make contact with the chloride resulting in a concentrated solution.

As the process continues a pressure builds up at the site of the salt due to the increase in volume and when it becomes high enough to overcome the adhesion of the coating a blister will develop. Blister formation slows down with time as the salt concentration decreases and the osmotic pressure becomes less. The pressure that develops inside the blister due to the increased water ingress can cause the blister to break.

Qualitative Tests

- A Qualitative test indicates whether a contaminant is present but not the exact amount.

Potassium Ferricyanide

This test indicates whether a contaminant is present but does not show the exact amount. Potassium ferricyanide is used to detect the presence of ferrous ions. In the test a fine mist of water is sprayed onto a small area of the blast cleaned surface and allowed to evaporate.

The test paper previously soaked in the potassium ferricyanide solution and allowed to dry is pressed onto the surface for 2 to 5 seconds. Any soluble salts present will react with the ferricyanide to produce a Prussian blue complex appearing as blue dots on the test paper.

Merckoquant Test

Test strips are available to determine semi-quantitatively many ions including chlorides, iron, nitrates and sulphates. The strip is dipped into the extracted liquid, the excess water shaken off and the colour produced compared with the scale provided. Different test strips are used for different ions and the deeper the colour the higher the concentration.

Quantitative Tests

- A Quantitative test gives an indication of how much contamination is present.

The test most widely used to assess soluble salt contamination is the Bresle test and is described in the **ISO 8502-6 "Preparation of Steel Substrates before Application of Paints and Related Products – Tests for the Assessment of Surface Cleanliness – Extraction of Soluble Contaminants for Analysis – The Bresle Method."**

The ISO standard describes the method for extracting soluble contaminants from a blast cleaned substrate using an adhesive patch.

Procedure for the Detection of Soluble Salts: ISO 8502-6

- The patch is attached to the substrate and then distilled water is injected into the outer edge of the patch by hypodermic syringe at an angle of about 30°.
- After a suitable period of time, to be agreed between interested parties, the solvent is sucked back into the syringe.
- The solvent is sucked back and forth at least four times without removing the syringe from the patch.

Note: On unpitted blast cleaned substrates a period of ten minutes has been found satisfactory as by this time 90% of the soluble salts have usually been dissolved. Then transfer the solvent back into the cup for analysis.

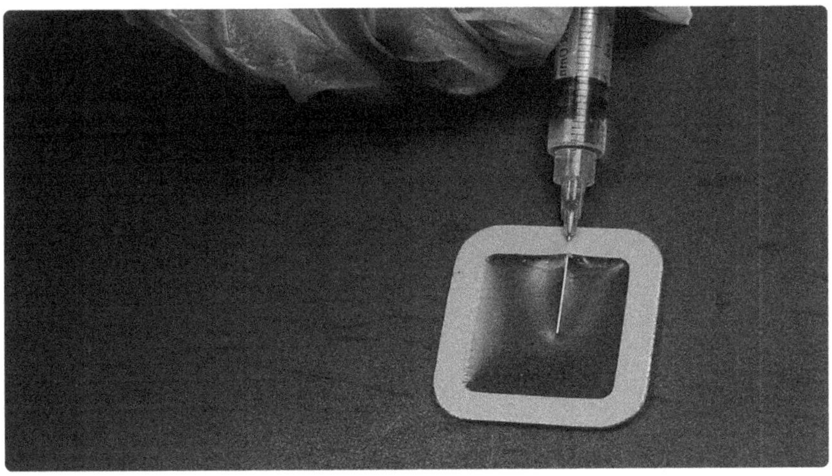

The Bresle Method ISO 8502-6

Any water-soluble salts found on the substrate are dissolved by the Bresle method. The conductivity is then measured with a conductivity meter to ISO 8502–9 Preparation of Steel Substrates before **Application of Paints and Related products – Test for the Assessment of Surface Cleanliness – Field Method for Conductometric Determination of Water-Soluble Salts.**

The conductivity of the water is measured using the conductivity meter and the total surface density of the salts within the test area is calculated by equation. Each project has its own pass/fail criteria, although normally within the offshore environment the salt contamination content should not exceed 20 mg/m^2.

Important: When performing the Bresle test method not only sodium chloride is dissolved as other salts are present on the surface. Because it is impossible to predict what other types of salts are present on the surface an "assumption" is made with the Bresle method.

The term "measured as sodium chloride" indicates that this mixture of salts is interpreted as "sodium chloride". The ISO 8502–9 method is calculated as a mixture of salts but expressed as "sodium chloride".

Additional Quantitative Tests:

- **ISO 8502-2** – Laboratory determination of chloride on cleaned surface
- **ISO 8502-5** – Measurement of chloride by the ion detection tube method
- **ISO 8502-9** – Field method for the conductometric determination of water-soluble salts
- **ISO 8502-10** – Field method for the titrimetric determination of water-soluble chloride
- **ISO 8502-11** – Field method for the turbidimetric determination of water-soluble sulphate
- **ISO 8502-12** – Field method for the titrimetric determination of water-soluble ferrous ions.

CHAPTER SIX
Environmental Conditions

Relative Humidity

Water vapour present in the ambient air can condense on the surface being coated. On clean metal surfaces condensation occurs when relative humidity reaches 100% (when the temperature falls below the dew point). Surfaces containing contaminants such as water-soluble salts can accelerate the corrosion process. Relative humidity (RH) is defined as the amount of moisture (water vapour) in the air compared with the maximum possible in the air.

If the RH reaches 100% then the air will not support any more water vapour and a surplus would appear as condensation. For that reason most coating applications require that coatings not be carried out if the RH exceeds a certain limit, usually 85% or 90%. In addition, if air is saturated with water vapour, the evaporation of solvent is inhibited, causing problems with film formation.

In practical terms the corrosion reaction starts on blast-cleaned steel surface when relative humidity reaches 60% – 70%. Relative humidity must be as low as possible during blast cleaning and painting activities. Coatings must be applied immediately after blast cleaning so the surface remains acceptable for painting and before any rust blooming or discolouration can occur.

Dew Point

Dew point temperature is defined as the temperature at which moisture will condensate on a surface. If the ambient temperature falls below the dew point, or if some or all of the structure has a temperature below the dew point, then condensation will occur. The temperature of the surface being coated should exceed the dew point of ambient air by at least 3°C immediately before and during application and during curing unless otherwise specified by the manufacturer.

If blast cleaning is performed when environmental conditions are close to the dew point temperature, condensation is likely and "flash rusting" may occur. The use of a dew point meter will make sure that the painting works can be executed if the ambient temperature, relative humidity and surface temperature are known and with limitations.

- **Condensation**: The presence of water in the form of condensation on cleaned surfaces can affect the adhesion and performance of any paint coating applied.

Many factors have an influence on the possibility of condensation including heat conduction of the structure, solar radiation on the surface, air flow around the structure and the presence of hygroscopic substances on the surface. It has for some time been recognised that the steel surface temperature generally should be 3°C above the dew point before painting is commenced unless the paints are described as being moisture tolerant.

Using a suitable instrument to measure air humidity and both air and surface temperatures the dew point can be determined. The minimum surface temperature (above the dew point) that is needed to avoid condensation can then be estimated. Many coating failures occur due to applying coatings over a damp or wet surface. The coating may look fine during application and even curing; however, if there was moisture present on the surface or in between coats there is a real possibility of delamination or adhesion failure.

A standard that gives guidance on the estimation of the probability of condensation is **ISO 8502-4 2009 (BS7079 Part B4: 1993) "Preparation of Steel Substrates before Application of Paints and Related Products Part B4 Guide to the Estimation of the Probability of Condensation on a Surface to be Painted."**

Environmental Inspection Equipment

Whirling Hygrometer

The whirling hygrometer or sling psychrometer consists of two thermometers mounted on the frame that can be whirled in the air by hand. The bulb of one of the thermometers is covered with a muslin sack and is wetted with water; this is known as the "wet bulb" – the other thermometer is known as the "dry bulb". The psychrometer is whirled by hand to force air past the bulbs, the dry bulb indicates the air temperature, the wet bulb helps determine the relative humidity.

Whirling Hygrometer

When the psychrometer is whirled through the air, water from the muslin cloth evaporates, and the evaporating water cools the wet bulb. The amount of cooling that occurs depends on the relative humidity, the lower the humidity the faster the water in the muslin cloth will evaporate and the more the bulb will cool. High humidity will cause less evaporation slowing the cooling process.

In air that has less than 100% relative humidity the wet bulb will record a lower temperature than the dry bulb; the difference in temperature is known as the "wet bulb depression". A special chart called a hygrometric table is used to convert the wet bulb depression to relative humidity or the humidity can be determined by a sliding rule called a Casella scale or a dew point calculator.

Electronic Dew Check Meter

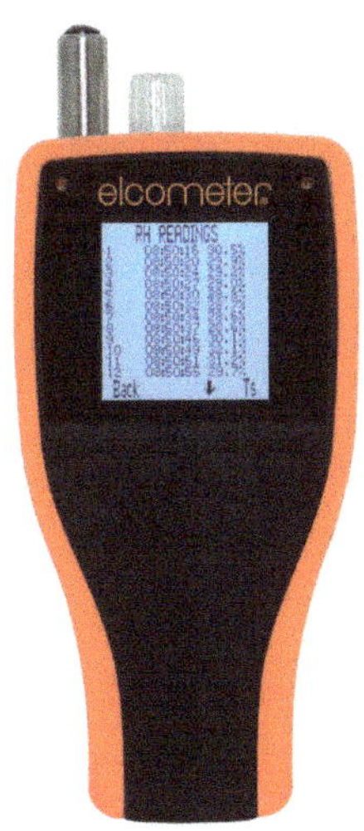

Electronic Dew Check Meter

The electronic dew check meter measures and records all climatic parameters required to treat surfaces. These gauges measure the relative humidity, ambient temperature, surface temperature and dew point temperature, and the difference between the surface temperature and dew point temperature. All of these parameters can be stored in the memory of the dew check meter and all data can be downloaded onto the user's laptop or PC. This type of equipment may not be allowed in certain environments (gas plants, offshore platforms) as it is powered by batteries.

Magnetic Steel Temperature Gauge

The steel temperature gauge or limpet gauge is simple to use and requires no batteries. These gauges are held to the surface by magnetic force (provided the substrate is magnetic). The gauge is placed on the substrate and left for 15 minutes until a constant temperature is achieved and the temperature recorded. It is important to remember where the gauge was placed as it could be painted over by the paint sprayer.

Magnetic Steel Temperature Gauge

Digital Thermometer

The digital thermometer is easy to operate. Simply open and place the probe on the surface for 20–30 seconds to take the reading and then record the surface temperature.

The pivoting probe and swivel tip ensure perfect surface contact and provide a highly accurate measurement. It is important that the probe is not dragged over the surface. These gauges are designed for industrial use so therefore contain no buttons or cables and provide a fold away action when not in use. If the gauge is to be used on hot substrates, a silicone sleeve can be attached over the probe and it also aids drop protection. These gauges may not be intrinsically safe as they require batteries.

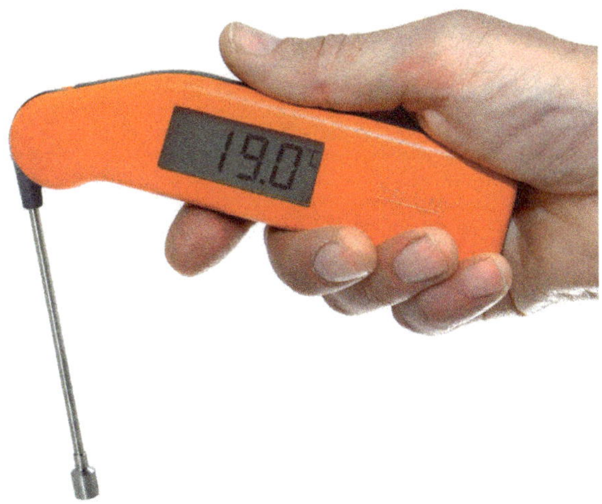

Digital Thermometer

CHAPTER SEVEN
Coating Materials

Paints have many different ways of working. They can passivate the surface, keep out air and water and impede the ion movement by creating a barrier effect. They can also "sacrifice" themselves by providing cathodic protection of the steel. The different coats in a paint system serve different purposes. Normally the paint system is built from a primer coat, an intermediate coat and a final coat. The function of the primer is to ensure good mechanical adhesion to the substrate and must be compatible with the subsequent paint system.

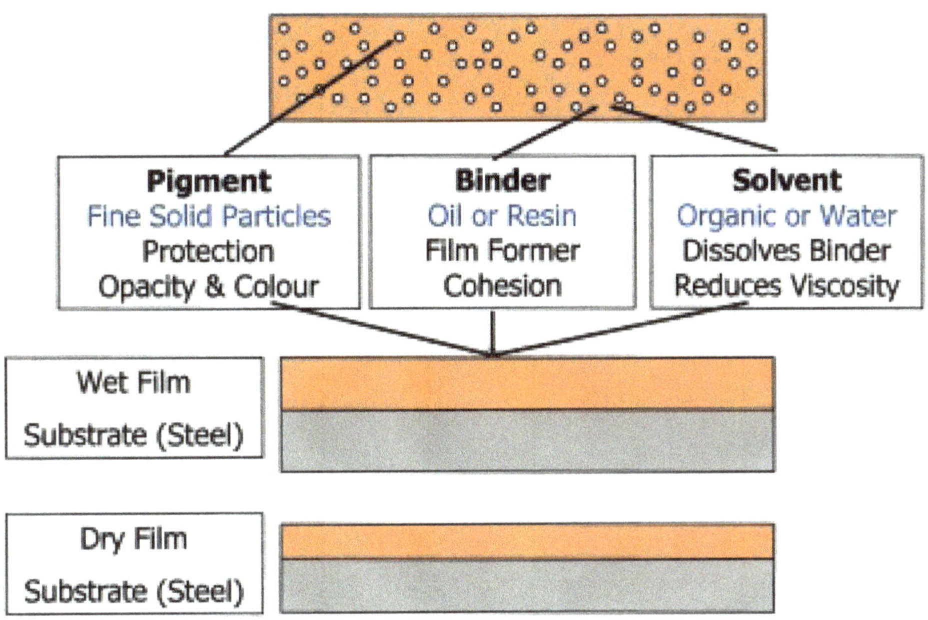

Constituents of a Paint

In a passivating paint system the primer coat must contain pigments to provide the passivating effect. The designation of such primers is often based on the pigment used. Passivating and cathodically protective primers must be in direct contact with the substrate in order to have the desired effect. The function of the intermediate coat is to build up the thickness of the paint system. This coat must also ensure that the finishing coat obtains good adhesion and is free from any paint film defects.

The final coat must provide the surface with the required colour and gloss as this is the coat primarily exposed to the sun, rain, and chemicals. The paint type of the intermediate coat is usually named after the binder, for example vinyl or epoxy. Normally the coating system is built up of similar based paint types in the intermediate coat and final coat. The primer coat may often be of a different type, provided it is compatible with the other coats.

Binder

The function of the binder is to ensure that the pigment particles are bound together and to ensure that the paint has proper adhesion to the substrate. To a large extent the binder determines the properties of the paint. In most paint types the binder is an organic material (polymer). In some cases the binder is inorganic such as silicates. The binder also determines the film forming process of the paint.

A distinction is made between physically drying paints which dry by evaporation of solvents and chemically cured paints which cure by two components reacting with one another. In a chemically curing paint the two components A and B react with one another while the solvents evaporate. A special variety of chemically cured paints is seen in the case of paints where the binder reacts with substances in the ambient environment, for example with the oxygen in the air or with water.

Oxidatively curing is the term used for the reaction with oxygen during the drying of an alkyd or an oil-based paint. Most chemically curing paints first undergo a phase of physical drying by evaporation of the solvents. It is important to note that paint which has only dried physically may be dissolved by the same solvents that were used in the paint.

Consequently, these paints have very good intercoat adhesion. The solvents in the coat dissolve the previous coat in such a way that the coats "melt" together. Their resistance to solvents and hydrocarbons is poor where when chemically cured, coatings are resistant to such chemicals.

Paints based on plastic dispersions, also called latex, are physically drying. In such paints the binder is finely dispersed as small globules in water. When the water evaporates the globules "melt" together to form a film. After this the paint film is no longer soluble in water, but it may be quite sensitive during the initial period.

Many of the binders are what is termed high-polymeric substances. This applies to latex binders as well as acrylic and vinyl binders dissolved in solvents. The high-polymeric substances are composed of long chains of similar molecule units. If only one type of molecule is repeated in this chain this is called a homopolymer.

If two different molecule types are incorporated it is called a copolymer. For example, a vinyl-acrylic copolymer consists of chains with alternating vinyl chloride and acrylate units. The high-polymeric substances consisting of long molecule chains are "thermoplastic". This means that they soften at increasing temperature.

High polymeric substances may also be built from large molecules in a three-dimensional network. Such networks are found in two component binders like epoxy and polyurethane, these are called "chemically cured resins". They do not soften like thermoplastic materials at increasing temperatures.

Pigments

Usually the function of the pigment is to provide colour and opacity (hiding power). They protect the binder from being broken down by the ultraviolet rays of the sun or actinic rays. Some pigments provide active corrosion prevention. They are primarily used in primer coats. Examples of such pigments are zinc chromate which has now been replaced by zinc phosphate due to the health hazard.

Zinc powder is used in paints providing cathodic protection. There must be metallic contact between the individual particles of zinc and between them and the steel surface in order for the paint to have its desired effect. Other metals like aluminium are used in the shape of pigment flakes. They strengthen the barrier effect of the paint by impeding the transport of air and moisture through the paint film. At the same time they have a reinforcing effect. Micaceous iron oxide (MIO) is also used to obtain a barrier effect.

The penetration of water both in its liquid form and as water vapour is important for the barrier effect. The vapour penetration is called "water vapour permeability". The amount of water vapour permeating through a paint sample during a certain period of time is measured at a specific difference in the vapour pressure on either side of the paint film.

Ordinary pigments are available as fine-grained powders. The particles are approximately 1μm in diameter and in some cases they are much smaller. During paint production it is important that the pigments are finely dispersed in the binder phase. To the greatest possible extent the pigment particles should be fully enclosed by binder and not form agglomerates together with other pigment particles.

Fine Grain Pigments

The process of having the pigments finely distributed in the binder is called dispersion or grinding. The pigments determine the colour, hiding power and gloss of the paint. The most important white pigment is titanium dioxide. The refractive index of titanium dioxide is higher than for other white pigments; this results in high opacity.

The opacity is determined by the difference in refractive index between the pigment and the medium around the pigment particles (the binder). Colour pigments can roughly be divided into inorganic and organic pigments. Within the group of inorganic pigments some are made of coloured minerals also known as "earthy colours".

Other inorganic pigments may be produced synthetically like red iron oxide and yellow iron oxide. The organic pigments provide the widest range of colours. These pigments are produced through a series of chemical synthesis processes. The basis may be raw materials from mineral oil or coal. In simplified terms the organic pigments can be said to have inferior resistance to, for example, light and heat than the inorganic types.

The Shape of Pigments is an Important Characteristic and can Influence the Physical Properties of the Coating. Some Common Shapes are:

- Nodular pigments are lump-shaped and generally add colour (such as titanium dioxide) or are sacrificial (such as metallic zinc). In practice most commonly used pigments are nodular.
- Acicular (needle-shaped). Acicular pigments, such as zinc oxide or glass fibres are used to reinforce and strengthen the coating film as well as provide colour. In the past asbestos fibres were often added to coatings to provide reinforcement (cohesion) and add some thixotropy. Although no longer used, asbestos pigments may be discovered when existing coatings are removed during maintenance painting.
- Lamellar, or plate-like, pigments, such as mica, aluminium flake, glass flake, and micaceous iron oxide, overlap when the coating dries, increasing the degree to which the coating is impermeable to moisture.

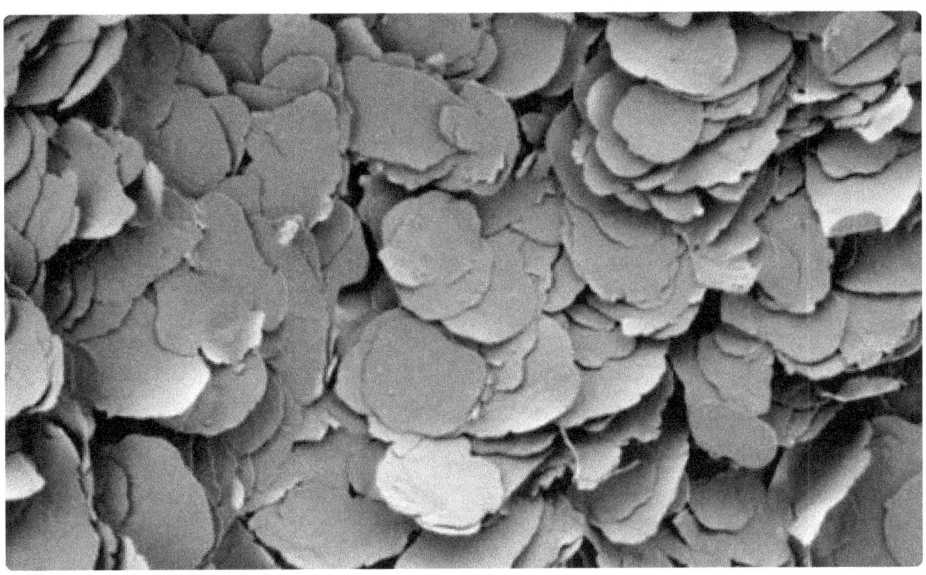

Lamellar Pigments

Solvents

Solvents in paint should be avoided, but binders often have properties which make the use of solvents necessary. The solvents dissolve the binder in such a way that the paint obtains a suitable viscosity for application. Some binders are solid and others highly viscous so that they are not suitable for use in a paint without a solvent. Only some chemically curing binders have such a low viscosity that no solvent is required. The solvent must be compatible with the binder as not all solvents dissolve all binders.

The paint factory chemist must know what solvent can be used. The painter must be careful to use the correct thinner if the paint is to be thinned down. The thinner is a mixture of solvents suitable for the paint; for epoxy paint a different type of thinner is used than for a vinyl paint. When paint producers are to select solvents they must not only take technical properties like dissolving power and evaporation rate into consideration but also factors such as health hazards and price.

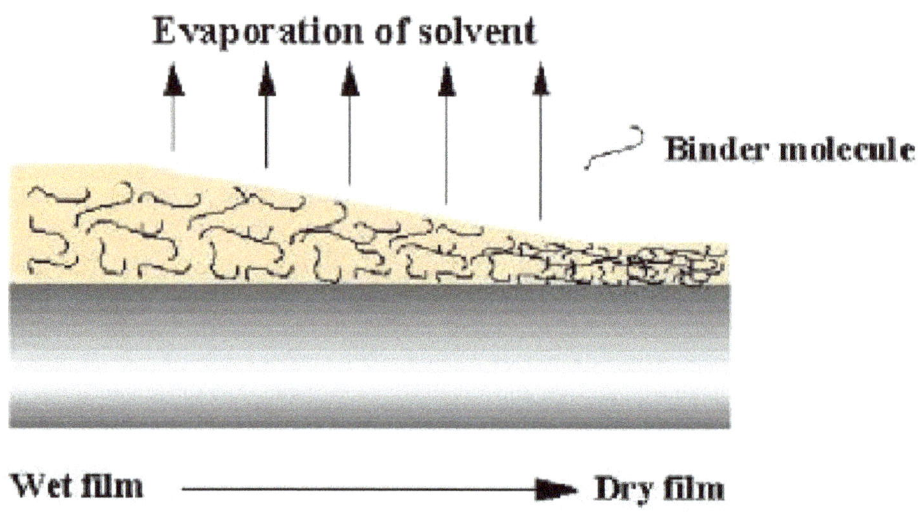

Solvents are volatile substances which means that they evaporate from the paint film more or less quickly. The boiling point provides an indication of the evaporation rate of the solvent – a high boiling point means slow evaporation. The solvent mixture used in a paint must have the correct rate of evaporation – if the solvent evaporates too quickly the paint will "set" too soon and be difficult to apply.

If it evaporates too slowly solvent entrapment in the film may occur, resulting in paint film defects in the coating. A much-used term is VOC. This stands for Volatile Organic Compound. This corresponds to the solvent contents and any other substances which may evaporate from the paint film. Several countries have imposed guidelines or restrictions in order to lower the VOC contents in paint. The objective is to reduce the environmental impact.

Extenders

Extenders, also called "fillers", are mineral materials used in finely powdered form in the paint. The extenders should provide body to the paint without impairing its properties. The extenders are usually white or light grey but in general they do not contribute to the opacity nor colour of the paint. A number of paint properties are determined by the ratio between the pigments or extenders and the binder in the paint film. An important issue in this respect is the "pigment volume concentration" (PVC).

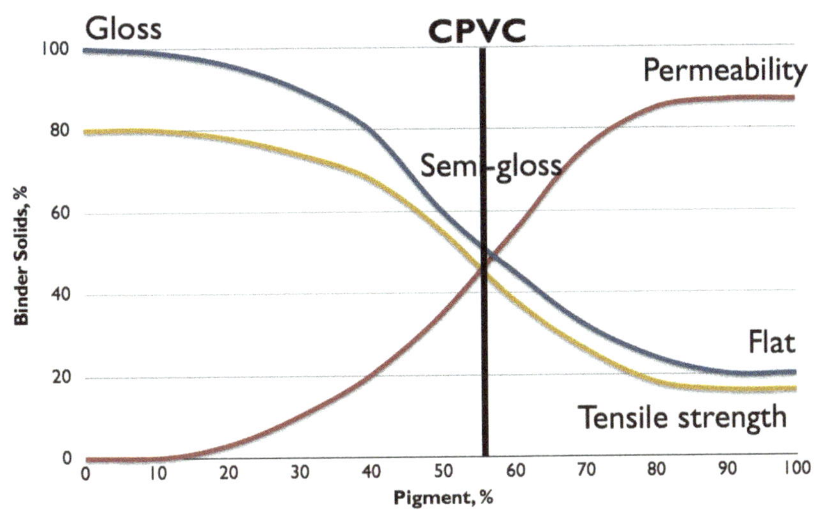

Levels of Pigmentation

PVC indicates the proportion of pigments and extenders in the dried paint film. With a high PVC the paint film has a low gloss and is more open to water and chemicals. It is primarily the engineer in the paint factory who calculates the PVC of the paint. In that case he or she calculates the volume of pigments and extenders contained in a litre of paint. This is divided by the volume of the binder plus pigments and extenders. The latter figure is the solids volume of the paint, the parts remaining after the solvent has evaporated.

Function of a Coating System

Primer

The primer is the most important coat of the entire coating system. When working offshore or in an aggressive environment a two-component epoxy primer is often used with a Bisphenol A or Bisphenol F resin make up.

Amine or polyamide curing agents (hardeners) are used with epoxy resin coatings.

The primer is responsible for:

- Keeping the substrate in its original condition
- Provides a key for the next coat of paint – Intercoat bond
- Adhesion to the substrate
- Adhesion to the intermediate coat
- Good chemical resistance.

When applying the primer coat it must be applied as a continuous wet film, free from any paint defects such as holidays and pinholes as moisture and condensation can migrate through these voids and contact the substrate leading to oxidation. When primers are applied on projects throughout the world, sometimes they are left to stand for weeks if not months – that's why they have to be highly adherent to the surface.

Primers provide two key elements: Adhesion to the substrate and good adhesion for the intermediate coats, a good intercoat bond is essential. The primer coat must be also be compatible for the intermediate and top coats and must thoroughly wet the surface. For corroded, uneven and heavily pitted steel work, the primer coat is usually applied by brush.

Paints containing corrosion preventive pigments are normally used for the primer coat. They may work by passivating the steel or they provide cathodic protection. If the paint does not contain such pigments it must be able to provide a good barrier effect. Several of the old pigments have gone out of use or are on their way out due to their toxicity. This is the case with red lead and zinc chromate. Zinc phosphate is an inhibiting pigment which has become more widely used in the protective coatings industry.

Organic Zinc Rich Primer

Zinc rich coatings are extensively used in the coatings industry. There are two main types of zinc rich paints used. The first with an organic binder and the second with an inorganic binder. The zinc contents must be at least 80% by weight of the solids content in both cases. Zinc rich paints are primarily used as primers; however, in certain environments they are also used without any further protection.

Organic zinc rich primers have some important characteristics when compared to inorganic zinc coatings. The most durable types have epoxy or polyurethane binders. In order to avoid the formation of of zinc soaps (saponification) it is important that the binder is alkali resistant.

Organic zinc rich coatings do provide limited cathodic protection as long as the formulation such as the particle-to-particle zinc content is maintained. The organic zinc rich coatings provide less cathodic protection than zinc ethyl silicate paint (inorganic). This is due to fact that the metallic content between the individual zinc particles is not as good and that the binder is not electrically conductive.

A binder in an organic zinc rich primer may be more or less chemical resistant depending on the binder and the environment it is used in. During the 1970s organic zinc rich epoxy primers were used in the marine industry and the offshore oil industry. These paints were often over coated with chlorinated rubber; however, zinc epoxy is most commonly used these days over coated with a high build epoxy. It is often claimed that organic zinc rich primers are less subject to critical surface preparation than are the inorganic zinc materials.

This may be true for the initial application since they would be subject to problems from organic contamination. On the other hand, eliminating the organic contamination factor, light rust colouration on the steel surface to be coated may be more easily tolerated by an inorganic zinc coating than by an organic based material.

This is due to the ability of the inorganic zinc vehicle to thoroughly wet the oxide and to react with it. All zinc rich paints require thorough surface preparation to Sa 2.5. This surface preparation results in a well cleaned and rough surface, which ensures that the paint adheres firmly to the steel. A hot dipped galvanized surface will require a light sweep blast to provide a key.

Advantages of Organic Zinc Primers

- Ease of application under a wide variety of environments.
- Ease of top coating without solvent bubbling, blistering or delamination.
- Can be applied in excess of 100μm without mudcracking.
- Resistant to splash and spillage of both fresh and salt water.
- Near equal performance to inorganic zincs when used as part of a two or three coat system with epoxy or polyurethane top coats.

Disadvantages of Organic Zinc Primers

- Not recommended for direct splash, spillage, or fume from inorganic acids, oxidizing agents, and organic acids.
- Very limited resistance to aromatic solvents, limited resistance to aliphatic solvents.
- Temperature resistance ranges from 120°C for phenoxy and epoxy versions to 232°C for silicone alkyd versions and 400°C for silicone versions.

Inorganic Zinc Rich Primer/Coatings

Zinc ethyl silicate paints are normally two component products where the zinc powder and the binder are supplied in separate containers. Before use the components are mixed in the correct proportions. Zinc ethyl silicate paints are based on an inorganic binder, ethyl silicate, which is electrically conductive. In order to obtain cathodic protection, it is therefore not of critical importance in this case that the zinc pigments are in direct contact with each other.

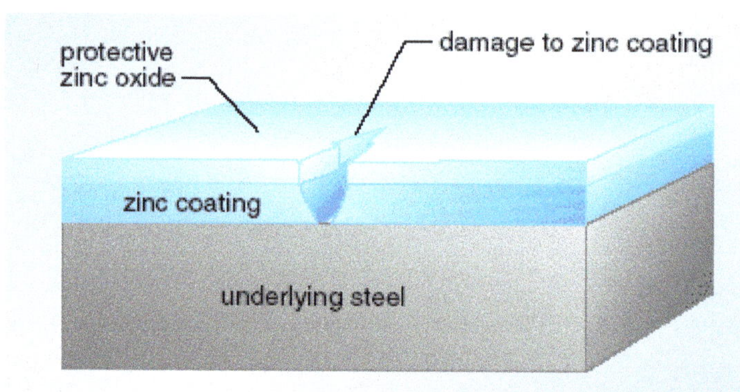

Example of a Sacrificial Coating

Zinc ethyl silicate provides excellent corrosion protection of the steel. Zinc ethyl silicate is generally used as a primer coat to be over coated with barrier paints. In the offshore industry epoxy paint systems are widely used on top of zinc ethyl silicate as far as structural parts above the water line are concerned. In the case of damage down to the steel, the zinc has a protective effect on the steel and sacrifices itself; however, an alkaline environment is formed around the damaged area.

It is therefore important that the paint system chosen is able to tolerate this environment. In the case of epoxy paints this is not a problem, but blistering may occur with alkyd and oil-based paints. The paint requires good electrical contact to the substrate in order to obtain the desired cathodic effect; the surface preparation is of particular importance. The cathodic effect is not obtained if the paint is applied on top of organic coatings.

Zinc Ethyl Silicate Powder

The paint requires direct contact with the steel in order to perform properly. The surface preparation required is abrasive blast cleaning to a degree of cleanliness of Sa 2.5/Sa3 with an anchor pattern of between 50μm and 75μm. Airless spray application provides the best results as it is not easy to apply zinc ethyl silicate by brush. The paint requires humidity to cure.

Too low humidity during the curing process results in low strength in the paint; this is also the case if the paint is applied in too-thick layers. The presence of humidity can take place in several ways. Some major works have large halls where temperature and humidity can be controlled. It has also been standard practice to sprinkle water on the constructions and if necessary cover them up in plastic to retain the humidity.

Prior to over-coating with organic coatings the inorganic paint must be completely cured. This is best controlled by the use of the solvent methyl ethyl ketone (MEK). The test method is available as a standard ASTM D 4752–87. In some cases sprinkling with cold water has resulted in imperfect through curing, not revealed by the MEK test. If the zinc silicate is applied too thickly the paint will fail and crack.

This type of cracking is called "mud cracking" because the surface of the coating looks like dried clay. Normally zinc ethyl silicate is specified at a dry film thickness of 75μm. The prevention of undercutting of a coating on steel is also an important anticorrosive property. Organic coatings generally do not have it. Inorganic zinc rich coatings do prevent rust from undercutting.

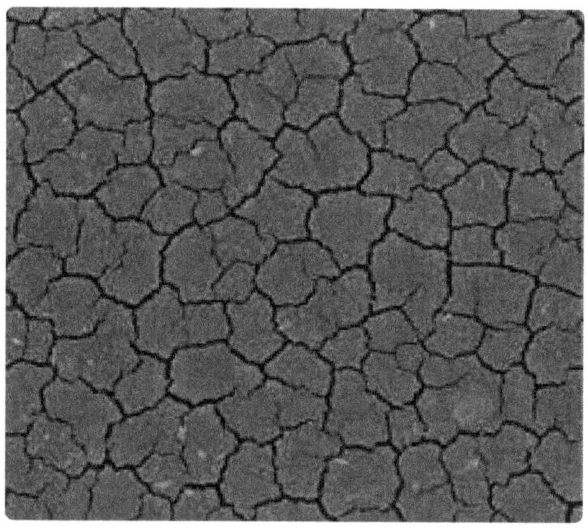

Mud Cracking

Both galvanizing and zinc silicate coatings have this property. The galvanized coating is amalgamated with the iron. Zinc silicate coatings provide this adhesion by the inorganic binder chemically reacting with the steel surface in a similar way to its reaction with the surface of the zinc particles. This reaction occurs at the interface between the steel and the coating, forming a permanent chemical bond between the two.

Protective Properties of Zinc Primers

Inorganic zinc pigments are slightly higher on the galvanic scale than carbon steel, therefore when placed into contact with each other, whether by spray application, galvanizing or metallization, they become anodic to carbon steel. The bimetallic connection provides sacrificial protection to the carbon steel substrate. To be truly sacrificial there must be point-to-point contact between the zinc particles and the carbon steel substrate.

The amount of zinc by weight in each litre of a zinc primer also has a direct relationship to the amount of sacrificial protection, particularly as the zinc primer ages. A considerable portion of the film is void (porous); however, this is not a problem due to the weathering process.

Any voids are filled in with zinc salts or white rust and this provides a barrier. The efficiency of inorganic zinc products in providing cathodic protection is directly related to their zinc loading. Products with the highest levels of cathodic protection typically have more than 90% zinc in the dry film. Less expensive formulations with lower zinc levels are available. Some products also use a less expensive filler material in place of the zinc that is still conductive but reduces the total life of the coating.

Top Coating Inorganic Zinc Rich Primers:

Problems can arise when top coating inorganic zinc rich primers. These can be categorized below:

Top coating too early

Different zinc rich primers require different levels of cure prior to top coating. The applicator should always use the most conservative data on top coating. Even under the best conditions the application of a mist coat is recommended in order to seal the porosities of the inorganic zinc film prior to application of the full wet coat, otherwise the top coat is prone to bubbling as the air and solvents evaporate from zinc film.

Even if bubbling does not occur, top coating too soon will result in the delamination at a later date when the top coat is stressed beyond the cohesive strength of the inorganic zinc film. The top coat that peels off will have a light layer of zinc on the backside of its film and the substrate will have the residual layer of the original thickness of the inorganic zinc.

Dry Spray

Dry spray is always a potential problem with inorganic zinc primers. These products are formulated beyond their critical pigment volume concentration (CPVC) in order to give the best possible cathodic protection from particle to particle contact between the zinc pigments and the steel substrate. Care must be taken to avoid adding too much solvent or too fast evaporating solvent.

Applicators must keep their guns parallel to the substrate approximately 9 to 12 inches from the substrate in order to achieve a wet film. Otherwise the applied film will be dry and uneven, resulting in a very porous dry film that resembles sand paper. The dry spray must be removed by sanding to avoid serious bubbling of the top coat.

Advantages of Inorganic Zinc Primers:

- Superior corrosion resistance in aggressive conditions.
- Very good long term performance as a single coat system in weathering environments.
- Excellent resistance to undercutting/rust creep under organic top coats.
- Temperature resistance up to 400°C alone and up to 593°C with silicone top coats.
- Excellent solvent resistance in a broad range of organic solvents.
- Good creep resistance on friction connections.
- Very abrasion resistant.
- Will not burn or support combustion.

Disadvantages of Inorganic Zinc Primers:

- Requires greater skill to apply than organic zinc primers.
- Tends to mudcrack when applied in excess of 75μm.
- Tends to promote blistering and holidays when top-coated too quickly with organic coatings.
- Tends to absorb contaminants such as dust, chlorides, oil, and grease during fabrication.
- Limited service life in water immersion, particularly at temperatures in excess of 69°C.
- Not recommended for prolonged exposure to acids below pH 5.0 or alkalis above pH 10.

Solvent Rub Test

Zinc ethyl silicate paints are two-component paints consisting of zinc powder mixed into a liquid ethyl silicate part. When the paint reacts with high humidity, alcohol is liberated, and it is extremely important that the alcohol has dispersed completely before over-coating. In recent years the ASTM D 4752 test method has become widely used. The dry film thickness is checked before the test and a test area is washed with water to remove loosely adherent zinc dust. Below is the procedure for the test.

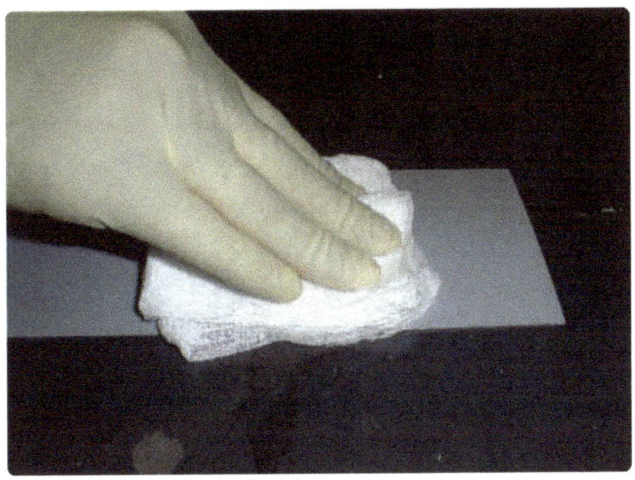

Measuring MEK resistance of Ethyl Silicate - Inorganic Zinc - Rich Primers by Solvent Rub (ASTM D 4752)

Procedure:

- Measure the film thickness and compare it to the specification.
- Remove all residues of loosely adhering zinc dust by washing with water before carrying out the test.
- Fold the cheesecloth into a pad containing four thicknesses of cloth and saturate it with methyl ethyl ketone.
- Rub the test area with the saturated cloth, exerting a moderate pressure with the thumb using a 50mm long stroke that encompasses the test area.
- Continue rubbing the surface with the MEK saturated pad, wetting the pad as necessary without lifting it from the surface until either the metal substrate is exposed or 50 double rubs have been completed.
- Select an adjacent area to be used for control. Repeat 1–5, except use a dry

cheesecloth to establish the effects of burnishing without the influence of MEK.

Resistance Scale:

5	No effect on the surface, no zinc on the cloth after 50 double rubs.
4	Burnished appearance in rubbed areas, slight amount of zinc on cloth after 50 double rubs.
3	Some marring and apparent depression on the film after 50 double rubs.
2	Heavy marring obvious depression in the film after 50 double rubs.
1	Heavy depression in the film but no actual penetration to the substrate after 50 double rubs.
0	Penetration to the substrate in 50 double rubs or less

Shop Primers

In terms of coating durability, surface preparation and paint work should preferably be performed once the final structure is complete. This is not possible in all cases and the costs of surface preparation can be very high. Therefore mill scale and rust are removed from steel plates or beams prior to the workshop stage. In such cases rust is removed using automatic centrifugal blasting equipment and the cleaning is inexpensive.

The steel surface is protected immediately after cleaning using a special paint material called the holding primer. The purpose of holding primer is to protect the cleaned steel surface temporarily during transport and the various manufacture phases until the final corrosion protection paint can be applied.

The prefabrication primer must have the following properties (ISO 12944–5, Annex B):

- The primer must form a uniform coat with high opaqueness.
- The primer must allow for the application of subsequent coats.
- The drying time of the primer must not be so long that the coated pieces cannot be handled shortly after application.
- The primer must protect the steel surface for a specified duration.
- The primer film must not significantly hamper welding or torch cutting operations.

The build up of the fumes and vapours from welding and cutting must not exceed the allowed exposure limits of the working site.

The function of the intermediate coat is to:

- Build up film thickness
- Resist chemicals
- Resist moisture vapour transfer
- Provide a key for the top coat.

Intermediate Coat

The main purpose of the intermediate coat or "mid coat" is to provide thickness for the complete coating system. The Intermediate coat provides strong adhesion to the primer coat and provides a key for the top coat. When applied in a corrosive environment like offshore, the intermediate coat provides a barrier to prevent aggressive chemicals and contaminants from penetrating the coating system and reaching the substrate. High performance coatings like glass flake epoxies or epoxies pigmented with aluminium flake are used.

The laminar pigments used in these high performance coatings proves a tortuous route for moisture or chemicals to pass through the coatings as they have to travel around each individual flake to reach the substrate.

The aluminium pigments provide a leafing effect like "fallen leaves on the ground" or "chinks of armour" and are widely known for their performance of protecting the binder from UV or actinic rays of the sun that leads to the degradation of the binder which in return generates to chalking. The plate-like pigments improve adhesion and reinforce the binder resin with the leafing pigment being far superior than its counterparts. Silicone alkyds are widely used on heat stacks that reach temperatures of up to 800 °C.

High Build Epoxy

High build epoxy coatings should be differentiated from high solids or 100% solid epoxy coatings in that they are usually coatings of the amine or polyamine type in which are added inert pigments. These high build materials are often very inert and are designed on the impervious coating concept, i.e. there are no inhibitive pigments included and the resistance of the coating is due to its completely inert characteristics.

These coatings are usually quite flat, have a semi-gloss or no gloss and are used under severe corrosion conditions. Many high build coatings are used as linings or they may be used alone or used in one or more coats. They can also be applied over inorganic zinc primers or as either a base coat or an intermediate coat which they will provide strength and thickness.

Top Coat

The functions of a Top Coat are to:

- Provide a barrier
- Resist UV light
- Resist against wind, rain, humidity
- Resist chemicals
- Provide a pleasing appearance – aesthetics
- Toughness
- Flexibility.

Top Coat

Final Coat

In the primer, intermediate coat and top coat system, the top coat provides a resinous seal over the intermediate coat and primer coat. The first top coat may penetrate into the intermediate coat thereby providing the coating system with an impervious top service. The top coat is the first line of defence against open exposure and condensation, highly polluted atmospheres, chemicals and the environment.

It forms a protective layer over the entire coating system and certain top coats like polyurethanes provide protection from UV and actinic rays. It impacts abrasion resistance as non-skid aggregate can be added to the top coat to provide a non slippery surface especially on offshore platforms where the environment is constantly wet and dry. Solar degradation occurs with amine based epoxy paints; however, polyurethanes have superior characteristics that prevent the binder from being degraded.

Adhesion

Types of Adhesion

Adhesion is defined "as the ability of a coating to adhere to a substrate". When surfaces are prepared by either blast cleaning or power tool cleaning the main objective of surface preparation is to create sound coating adhesion. Coating adhesion is the most critical part of any painting project – without proper adhesion the system is on course for premature paint failure.

The good adhesion of coatings is essential when applied in aggressive and submerged locations which are highly corrosive. The correct level and method of surface preparation is key for longevity of a coating system. Remember the motive for surface preparation is to ensure the maximum bond strength will advance at the interface between the substrate and the coating.

- Failure of a coating from the substrate – **Adhesion Failure**
- Failure within a layer of paint – **Cohesion Failure**
- Failure within the coating system itself (in-between layers) – **Delamination.**

This type of adhesion or bond strength relies on the surface being prepared and the coating system being applied. We will cover two classifications of adhesion:

- Chemical Adhesion
- Mechanical Adhesion.

Chemical Adhesion

When we talk about chemical adhesion we refer to the chemical bond which is generated between the substrate being prepared and the coating being applied. Two categories of coatings fall into chemical adhesion; these are hot dipped galvanized coatings and inorganic zinc coatings. The hot dip galvanized coating (zinc) alloys and consolidates within the steel substrate while immersed in a bath of hot molten zinc at temperatures of between 450°C and 475°C; this provides ideal chemical adhesion.

This bond is also referred to as a "metallurgical bond". Inorganic zinc coatings have a chemical bond between the zinc silicate molecules and the steel substrate. The main advantages of inorganic zinc coatings are that they do not undercut and are not susceptible to under film corrosion due to this chemical bond.

Mechanical Adhesion

Mechanical adhesion is associated with surface profile or anchor pattern. Surface profile or tooth is created by peaks and valleys on the substrate normally created during blast cleaning with a sharp cutting edge abrasive. Enhancing the surface area of the substrate will also increase the adhesion properties of a coating. This can be done by mechanical or physical means.

With physical means, the surface area can be increased by abrasive blast cleaning which produces a greater surface area to which the coating can bond to; this leads to greater adhesion. When applying a high solids coating, for example glass flake or a metallic coating as in thermal sprayed aluminium, a surface profile of between 75μm and 100 um is required for optimum adhesion. The lack of surface profile can increase the risk of adhesion failure as all coatings require some form of "surface profile or key".

Loss of Adhesion

Loss of adhesion is often due to the fact that one or more layers of paint in the system are easily removed from another coat previously applied or from the steel substrate. In the case of peeling, the coating comes off in large or small flakes. Loss of adhesion may occur due to insufficient surface preparation of the substrate. If a rusted surface is coated the paint will not adhere sufficiently to the surface.

If in this case the result is loss of adhesion, the flakes can be removed and in most cases remnants of rust or similar contaminants are often found on the back of the paint flake. Another cause may be that too long a time has elapsed from application of one coat to another. In this instance the over-coating interval has expired. Peeling often occurs if remains of dust and dirt have not been removed between application of coats.

Contamination between the coating layers may mean that the coating will not obtain the required adhesion to the paint coating beneath. If paint flakes are examined more closely we may see remains of contamination on the back of the paint flake, for example dust or dirt. In some cases loss of adhesion may occur if old but smooth coatings are recoated.

If the relative humidity has been too high during curing of epoxy coatings, the surface may appear sticky, viscous or somewhat discoloured. This may derive from amine exudation. Amine exudation may occur if too much carbon dioxide is present on the surface. It is important that areas with exudation or blooming are removed by fresh water washing or solvent cleaning as flaking could occur.

Thermoset and Thermoplastic

There are two main types of curing processes that occur.

- **Thermoset – Convertible (will not resoften)**
- **Thermoplastic – Non Convertible (will resoften).**

Thermosetting Coatings: Thermoset coatings cure by a chemical reaction. A chemical reaction occurs between the coating components which modify the coating chemistry. This reaction is called "polymerisation". The "cross linking" of the multiple components becomes the solid film. Cross links are bonds which hold the polymer chain together. Thermoset coatings have strong rigid crosslinks. If the paint's solvent is reapplied it will not resoften.

Thermoplastic Coatings: Thermoplastic coatings dry solely by solvent evaporation; the resin is already in its final form. There is no chemical or physical change in the non-volatile part of the coating that forms the film. Cross links in the thermoplastic process are weak and flexible. If the paint's solvent is reapplied it will resoften.

Physically Drying Paints

These paints dry solely by solvent evaporation. The binder originally existed in a solid state and it reverts to the solid state during drying. No chemical reaction takes place during the drying process. The evaporation of solvents is slower at low temperatures, but it does not stop. Provided that other conditions are acceptable, for example no ice on the surface, the paint can also be applied at sub-zero degrees.

The paints are reversible; when solvent is applied to these coatings they will resoften. When the same type of paint is used for over-coating, the underlying paint film will soften a little. The result is good adhesion between the old and new paint. If thorough degreasing has been carried out, maintenance painting on top of a physically dried paint will rarely present any problems.

Chlorinated Rubber

These paints are known as "thermoplastic" paints. The paint also contains plasticizers that help to provide the binder with good mechanical properties.

Advantages of Chlorinated Rubber Coatings

- Fast to dry and cure even in cold temperatures.
- Broad range of water, acid, and alkali resistance.
- Excellent re-coatability within multiple coat systems, seldom delaminates.
- Excellent cohesive strength, resists delamination.

Disadvantages of Chlorinated Rubber Coatings

- Tend to dry spray or cobweb particularly during hot dry weather.
- Poor solvent resistance.
- Requires three or more coats to achieve good corrosion resistance.
- Low volume solids do not meet current VOC restrictions.
- Limited heat resistance.
- Chlorinated rubber begins to disintegrate at 80 °C.

Vinyl Paints

Vinyl paints are closely related to chlorinated rubber paints. They dry as the solvents evaporate, and they are reversible. The vinyl binder is produced synthetically and may consist of different copolymers with polyvinyl chloride. It is soluble in aromatic hydrocarbons, ketones and esters. The solids content in vinyl paints is low and rarely above 30–35% by volume. Consequently, more coats must be applied in order to obtain the film thickness required.

Vinyl paints were at one time widely used in industrial environments and could withstand exposure in quite severely corrosive areas. They have good resistance to inorganic acids and alkalis. In addition, vinyl paints are relatively resistant to water, oil and alcohols. Vinyl paints have been used widely in the offshore industry. Due to VOC restrictions the use of vinyl paints has decreased over the years and has been superseded by epoxy coatings.

Advantages of Vinyl Coatings:

- Easy to apply and fast to dry even in cold weather.
- Excellent recoatability within multiple coat systems, seldom delaminate.
- Excellent water and corrosion resistance even in immersion.
- Excellent adhesion to most inorganic zinc primers.

Disadvantages of Vinyl Coatings:

- Poor adhesion direct to metal, requires special acid-modified primers.
- Low volume solids do not meet current VOC restrictions.
- Requires three or more coats to achieve corrosion resistance.
- Poor solvent resistance.

Acrylic Paints

These paints have several properties in common with chlorinated rubber and vinyl paints. They can be distinguished in particular by better sunlight resistance. As a result, painted areas will retain their colour and gloss better. The resistance to solvent exposure is slightly lower than chlorinated rubber and vinyl paints. Acrylic paints have essentially the same areas of use as chlorinated rubber and vinyl but are preferred where colour and gloss stability are of major importance.

Acrylic paints can be used as finishing coats on top of vinyl paints. Airless spray is the most suitable method of application whilst brush and roller are less desirable. Waterborne latexes have been around for a long time, but their limited resistance to aggressive chemicals and corrosive conditions have restricted their use mostly to aesthetic or architectural purposes.

Known for their exceptional resistance to ultraviolet rays of the sun the waterborne acrylics retain colour and gloss longer than most other top coats including polyurethane. The major shortcoming of single component waterborne acrylics from a protective coatings standpoint is their inability to cure at low temperatures which drastically restricts the painting season in cold and humid climates.

Advantages of Acrylic Coatings

- Ease of application by brush, roller and spray.
- Relatively inexpensive.
- Excellent colour and gloss retention.
- Long term flexibility.
- Good adhesion to galvanized surfaces.

Disadvantages of Acrylic Coatings

- Early flash rusting when applied in temperatures approaching freezing.
- Surface must be free from all residual oil and grease as they do not wet the substrate well.
- Slow cure during high relative humidity conditions.

Waterborne Coatings

Waterborne anti-corrosion paints based on different binders are available. The largest group is dispersion paints based on acrylic or a combination of acrylic and other substances. Such combinations are the result of reacting the substances with each other during the production of the dispersion. Copolymer compounds are produced, thereby certain properties can be improved compared to those of the substances individually.

The binder is a dispersion (small droplets or globules finely dispersed in water) with the size of these globules being less than 1µm. The reason why waterborne paints are so popular is due to the health hazard associated with solvents. This does not mean that the waterborne paints are without any issues. It must be taken into consideration that the waterborne anti-corrosion paints do contain some solvents which are to assist in the film formation.

Additionally, they contain biocides which are additives to prevent attacks from bacteria and fungi. Apart from that, pigments, inhibitors and other substances in the paints may have an effect that is harmful to health. The risk to being exposed to these substances depends on the method of application and the protective equipment used. During spray application spray mist is formed; the particle sizes of this are between 100µm and 400µm.

The part of the spray mist which can be inhaled and end up in the lung alveoli is likely to be much lower than 1% of the total sprayed paint. In addition to the mist, a continuous evaporation from the painted surfaces takes place. For health and safety reasons the waterborne anticorrosion paints based on acrylic or other binders must be considered to be beneficial due to less exposure to solvents.

When spray applying them, it is still necessary to wear the correct PPE and RPE to protect the applicator against inhalation of spray particles and to protect the skin. The exposure is lower for the people working nearby and who normally do not wear any form of protection. Paint application causes a risk of allergy from skin contact or damage due to solvents in the air. Consequently, the different requirements for protective precautions must be evaluated.

The binders of the most common waterborne paints are copolymers of styrene acrylate or vinyl chloride – vinylidene chloride acrylate. Some waterborne two-component epoxy paints have been available for use in anti-corrosive systems. Their mode of action may be galvanic or inhibitive like zinc phosphate or based on a barrier effect.

The waterborne acrylic and vinyl paints are physically drying paints; after the binder globules have fused together the binder will not dissolve or disperse in water again. The water was never a solvent for the binder, this is how waterborne dispersion paints are distinguished from other physically drying paints. For alkyd and epoxy paints an oxidative or chemical curing takes place after the water has evaporated.

The water in the paint produces "flash rust" if effective inhibitors are not used. They must be effective as long as the paint is liquid; as such inhibitors are water soluble. An example of inhibitors is sodium benzoate possibly mixed with sodium nitrate. There is a risk of flash rusting when the paint dries at a relative humidity of the air that is more than 65%.There is also a risk of these inhibitors being left on the surface and causing adhesion problems, and many specifications do not allow their use.

The permanent corrosion preventive effect is primarily based on special pigments in the primer coat. The paint must have a water uptake in order for the relatively water-soluble anti-corrosive pigments to work. The anti-corrosive pigments used include zinc phosphate, zinc borate, zinc molybdenum phosphate and aluminium triphosphate. Paints based on dispersions are usually not as diffusion tight as solvent based paints.

If the anti-corrosive effect is based on the barrier effect, the development of suitable paint formulations presents a number of problems. The pigment dispersion is particularly important. Consequently, factors such as the pigment contents (Pigment Volume Concentration) and the dispersion agent for the pigments are of major importance. The waterborne vinyl and acrylic paints can be produced in many colours and they are not easily disintegrated by sunlight.

They can therefore be expected to retain their colour and gloss quite well. On the other hand, the paints are not strong on areas with much condensation or areas immersed in water. The resistance to acidic industrial gases, salt containing air and a number of chemicals is average. The temperature of the substrate should preferably be below 60°C.

The areas of use for waterborne paints are primarily less aggressive environments but products have been seen to perform well in corrosivity category 4, provided that the paint system has been built up properly. The paints can be applied on zinc rich primers and on galvanized surfaces. Waterborne paints can be applied by brush, roller and airless spray. During application the temperature of the substrate should not be below 5°C.

Usually over-coating can take place after a couple of hours; however, details will be on the paint manufacturer's data sheet. It is important to note that the drying time depends on both the relative humidity, the air velocity across the surface and the temperature.

Since the paint is sensitive to humidity during the period before it is sufficiently dry, it is important to try to protect the construction and ensure optimum drying conditions for the first 24 hours after the paint has been applied. If the temperature is low and the humidity high in the ambient surroundings, good ventilation must be provided, otherwise it may be expedient to dehumidify the air around the structure.

Oxidatively Cured Paints

Some binders may react with oxygen in the air and thereby cure to a solid film. This can be described as a special type of chemical curing. The binder must be based on a content of drying oils such as linseed oil which contains unsaturated fatty acids. They have a tendency to grab hold of oxygen molecules in the air, react with them and thereby build a network of several fatty acid chains. Unsaturated fatty acids have been incorporated in alkyds, also cure by reacting with oxygen.

During the initial period after application of the paint, drying takes place as the solvent evaporates; after this the actual curing starts. Several factors have an influence on the curing process such as low temperatures prolonging the curing. This is also the case of limited air supply and extra high and extra low relative humidity of the air. In order to increase the curing rate metal compounds called driers are added.

Alkyds

Alkyds are among the oldest and the most basic of the paint forms. They are produced through a reaction which include various combinations of an alcohol, an acid and a fatty oil. The three ingredients are combined and react to create an alkyd. The term alkyd comes from the "al" in alcohol and the "cid" in acid. The ratio in which these products are combined, the functional groups they contain, and the additives control its ultimate properties.

The combination of the alcohols and the acids form what is called an "ester". This ester linkage is a weak point in the polymer when it is attacked under alkaline conditions (high pH). Alkyd coatings which can be created from naturally occurring materials, cure by oxidation. Oxidation creates linking of the resin along the fatty oil chains; this reaction produces the basic chemical structure.

In addition to the three basic ingredients, modifiers can be added to alkyd coatings to enhance performance. Driers are commonly added and are used to drive/control oxidation. These driers include cobalt, manganese, zirconium and calcium which promote a controlled reaction of the applied film so it dries and cures throughout the film, rather than the outer layer where oxygen is more readily available.

The Properties of an Alkyd Coating are Dependent on the Variables Below

- The type of drying oil fatty acid used to make the alkyd.
- The relative amount of drying oil fatty acid to synthetic resin components commonly referred to as the "oil length".
- Long oil alkyds contain 60% to 65% oil.
- Medium oil alkyds, more widely used as a protective coatings than either long or short, may contain between 30% and 60% oil.
- Short oil alkyds may contain less than 35% oil.
- The ingredients and the processing methods used during the manufacturing of the resin itself determines the properties of the synthetic resin component (polyester).

Types of Drying Oils and Alkyds

- **Fish Oil**: Fish oil (menhaden fish oil) dries more slowly than linseed oil, and the resultant film is softer than other types of oil. It has very good penetrating properties, which has led to its extensive use in surface tolerant alkyd primers.
- **Linseed Oil**: Linseed oil is used mainly in the preparation of air drying alkyd resins, epoxy ester resins and urethane oils or uralkyds. Linseed oil has good penetrating properties but poor colour stability.
- **Safflower Oil**: Obtained from the safflower seed, this oil is used to make high quality air drying alkyds and oil modified polyurethane resins.
- **Soya Oil**: This semi-drying oil is widely used in medium and long oil alkyds for architectural interior and exterior coatings. They air dry rapidly and have good colour retention and exterior durability.
- **Tall Oil**: Tall oil is a by-product from the manufacturer of paper from wood pulp and is a semi-drying oil used mainly in alkyds and epoxy esters as a low cost resin. Its durability on exterior exposure is less than that of soya oil.
- **Tung Oil**: A product of the tung tree, this oil contributes properties to the coating films that are tough, flexible, harder and more water and chemical resistant than vehicles based on other oils, such as linseed. This oil dries rather rapidly so it is commonly used in combination with other resins, such as phenolics, for bridge painting systems.

- **Castor Oil**: This semi-synthetic oil is used in the manufacture of high quality alkyd and epoxy ester resins with faster drying and better colour than soya based reins. Epoxy ester coatings based on castor oil provide excellent vehicles for primers in both air drying and baking systems. Castor oils are used in both air-drying and baking systems.
- **Styrenated Oils**: Reacting the fatty acids with either styrene or vinyl toluene results in an alkyd that is fast drying and tough with good water and alkali resistance, but reduced ultra violet and solvent resistance.

Modified Alkyds

One of the enduring qualities of alkyds is their compatibility with other resins, either in cold blends or "cooked processes", that result in improving the protection characteristics of both components.

Examples are:

Phenolic Alkyds: Modifying the alkyd resin with a phenolic resin results in a new resin that cures fast and has improved water and corrosion resistance. The downside of this combination is reduced UV resistance, making them chalk rapidly on exterior exposure. They are widely used in formulating shop primers and so called "universal primers" compatible with a wide range of top coats.

Epoxy Esters: By modifying the alkyd resin with an epoxy resin, a harder film and an increased resistance to splashing and spillage of hydrocarbon fuels and solvents is obtained. Adhesion is also usually better than unmodified alkyds. The downside is a reduced UV resistance, making them chalk rather rapidly.

Silicone Alkyds: The addition of silicone to the alkyd resin creates a win-win situation with only an increased cost as the downside. They have outstanding gloss retention and weathering resistance, as well as increased heat resistance, compared with alkyd resin alone. Although ordinary alkyds rarely have heat resistance greater than 121°C, silicone alkyd modifications readily achieve heat resistance in excess of 232°C. They also retain their flexibility better with age due to the silicone component.

Urethane Alkyds: The polyurethane resin component adds toughness to the film with increased UV and weathering resistance. One use of this technology is in floor enamels. They also have greater alkali resistance than traditional alkyds. Drying and curing times are improved at low temperatures.

High Solid Alkyds

Another result of increasing VOC restrictions has been the development of higher volume solid alkyd resins by the raw material manufacturers. Although many useful improvements have come from these newer resins, care should be taken in selection and use of the resulting paint or coating. Simply changing from a low solids alkyd resin to one with higher solids is not necessarily beneficial.

The resin manufacturing process requires very careful selection and balancing of solvent/resin solutions to maintain the same protection properties afforded by the older, lower solids versions. One of the potential problems that must be overcome is the tendency to apply high film thickness with higher solids versions. They still dry and cure by reaction with oxygen from the atmosphere, and the thicker the film, the greater the demand for oxygen to achieve a thorough cure down to the substrate.

Improperly formulated versions can result in soft or otherwise partially cured layers of coating of the substrate resulting in poor adhesion and corrosion prevention properties. They also contribute to wrinkling of the topcoat due to the stress of a hard surface film over a soft underlaying film.

Advantages of Alkyds

- Good weathering properties particularly on exterior structures.
- Easy to apply by brush, roller or spray. Good gloss retention, flow and levelling properties result in a smooth attractive film.
- Good wetting properties particularly long oil alkyds makes them applicable over marginally prepared substrates such as ISO 8501-1 Rust Grades C and D.
- Relatively inexpensive.
- Synthetic alkyds have very good gloss, whereas natural oils have very low gloss.

Disadvantages of Alkyds

- Relatively poor resistance to water, should not be used in immersion or frequent wet/dry cycles.

- Relatively weak alkali resistance, will form a soap and saponify when applied over inorganic zinc, zinc galvanizing and concrete.
- Slow drying and curing in low temperatures.
- Up to three coats are necessary to build up adequate film thickness.

Saponification

Special consideration should be given to surface preparation when using an alkyd coating. If the surface is not properly cleaned and alkali species remain on the surface a **Saponification** reaction can take place. Saponification occurs when a base reacts with a fatty oil and forms a soap. Saponification on the surface of a substrate before subsequent coating application can compromise coating adhesion.

When applying an alkyd ensure the film is not applied too thickly. The top of the film can cure more quickly than the bottom of the film near the surface. This is because oxygen cannot reach the bottom of the film to initiate cure which can cause adhesion and crosslinking problems. The surface of the alkyd coating will have the appearance of "wrinkling"; whilst the coating will be hard on the top of the coating, it will remain soft beneath skin of the coating.

Epoxy Esters

Epoxy esters are a type of alkyd in that a high molecular weight epoxy resin is reacted with one of the alkyd resins. This reaction is an important one because of the rather extensive use of epoxy ester coatings. This reaction that takes place is usually between solid grade epoxy resins and drying oil fatty acids. This reaction requires elevated temperatures to form the epoxy ester, therefore this reaction does not take place except under manufacturing plant conditions.

The end product, the epoxy ester, can then be used to make the air drying epoxy ester coatings. While the epoxy resin becomes a chemical part of the epoxy ester molecule, the drying or curing of the coating results from the oxidizing reactions of the unsaturated fatty acid part of the resin; it therefore dries like an oil based enamel.

Such a combination improves the chemical resistance of the basic polyester or alkyd resin and makes them a step above the ordinary alkyd resin for chemical resistance especially under alkaline conditions. Epoxy esters should not be confused with two component epoxy paints as the word "epoxy" does form part of the name; however, this is a one component type of coating. Epoxy ester paints provide a very hard coating and make a good machinery finish or similar types of products.

They have usable resistance for use outdoors; however, epoxy esters tend to "chalk". Chalking takes place when the binder degrades in the outer most layer of the paint due to UV attack by actinic rays, thereby pigments and fillers are exposed to the atmosphere. When the surface of the coating is gently rubbed with a piece of black felt or velvet, the chalk can be seen to come off onto the material. Below is a simple test to evaluate the degree of chalking.

ASTM D 4214-07 Standard Method of Evaluating Degree of Chalking of Exterior Paints

ASTM D4214–07 is simple to use:

- A piece of black felt or velvet is placed over the index finger or thumb.
- Rotating the finger at a 180° degree angle when pressed against the surface.
- Remove the cloth and evaluate the degree of chalking against the standard.
- To minimize chalking polyurethanes paints combined with chalk resistant pigments are used. Acrylic modified epoxies and with UV absorbers are applied where UV exposure is critical, especially in the Arabian Gulf Peninsula.

Silicone Paints

Silicone paints can be considered as a combination of glass (silica) and organic binders. The binder is expensive and is often modified with other binders such as alkyd or acrylic. Silicone alkyd has considerably better weatherability than both pure alkyds and other modified types. The heat resistance of the silicone paints is remarkable as some types withstand temperatures of up to 500°C while others may tolerate temperatures of 600°C or higher.

Silicone paints are very resistant to outdoor exposure and have less tendency to discolouration, chalking or loss of gloss. Silicone acrylic is also used for the formulation of heat resistant paints and are tolerant of temperatures of approx 250°C. Aluminium silicone is used for the sealing of thermally sprayed aluminium coatings (TSA) and is often referred to as the "sealer coat". TSA coatings are commonly used on exhaust pipes, flare booms and gas turbines.

Chemically Cured Coatings

Corrosion protection by means of two-component paints such as epoxy and polyurethane paints is obtained by barrier effect. Two component paints are delivered in two separate containers. The components are called "base" and "curing agent". The containers are often marked A (base) and B (curing agent). Before use the components must be mixed in the correct proportions. The components should always be mixed by using power tools such as mechanical mixers.

Part A (Base)/Part B (Curing Agent)

The mixing ratio is usually written on the side of the can or on the paint data sheet. The correct mixing ratio is extremely important otherwise problems may arise if too much or too little curing agent is added to the base. After mixing a chemical reaction takes place between the components. The chemical reaction is temperature dependent and is faster at high temperatures.

Consequently, the pot life (the time in which it can be used) of the paint after mixing decreases with increased temperature. Some types need to be left for a while after mixing (induction period) before they are used. As the paint cures, a film with a three dimensional and insoluble network is formed. These paint films are extremely strong and resistant to chemicals, acids and alkalis.

Epoxy Paints

Epoxy paints form a very strong and hard paint film. Pure epoxy paints are usually divided into solvent-free, high solids, solvent-borne or water emulsion paints. The base of epoxy paints consists of epoxy molecules of varying length. Solvent-free and high solids epoxy paints usually have low molecular epoxies with short molecular lengths.

Solvent-based epoxy paints with a solids volume of less than 60% usually contain high molecular epoxies and have longer molecule lengths. Generally these types require strong solvents. A number of different curing agents can be used in epoxy paints. The most common curing agents are polyamines, polyamides, amine adducts or isocyanates. The more common epoxy protective coating ares based on one or more of the following types of epoxy resins in the base portion of the product.

Bisphenol A: The most commonly used resin for epoxy coatings comes in a wide range of properties achieved by blending different molecular weights of the resin. The resins are usually very user friendly in their surface preparation requirements and application/curing properties. Depending on the formulation they can provide toughness and resistance to both chemicals and heat.

Blending with phenolic resins gives them even better hardness and temperature resistance. One high molecular weight variation of Bisphenol A is phenoxy. They have excellent adhesion properties with good flexibility and impact and adhesion resistance. They are quite often used in single pack, zinc rich epoxy primers.

Bisphenol F: These are generally more reactive than Bisphenol A resins, with some curing at low temperatures. These epoxies normally have better heat, solvent and chemical resistance than epoxies based on Bisphenol A resin. Most important they may be formulated well below the VOC threshold. Quite often epoxy products in the 90% to 100% solids range use Bisphenol F in the base portion.

Multifunctional Epoxies

These epoxies are much more effective than Bisphenol A or F epoxies and result in even greater chemical resistance, particularly as tank linings. Greater care must be taken in application and curing sometimes with heat to achieve the maximum density of the coating film. Control of film thickness is critical to avoid excessive stress in the film which can lead to delamination.

Speciality products using the base resin are capable of withstanding temperatures as high as 260°C and pressures as high as 10,000 psi. Many tank linings are based on the epoxy phenol novolacs so they may be identified in the product data sheets as either epoxy phenolics or novolac epoxies. When cured with amine curing agents these epoxies are very resistant to most solvents, mineral acids such as hydrochloric, hydrofluoric and sulphuric acid, and alkalis such as liquid ammonia.

Post cured systems are even resistant to immersion in methanol. Numerous versions of epoxy/phenolic/novolac products can be found in this group. This modification provides the thermal stability of the phenolic backbone with the reactivity and the versatility of the epoxy resin.

Coal Tar Epoxy Paints

Coal tar epoxy is a two-component epoxy paint modified with coal tar. The modification provides a less expensive paint. In addition, the coal tar contents provide the paint with flexibility and excellent water resistance. The epoxy imparts high resistance against chemicals and solvents. Coal tar epoxy contains no corrosion prevention pigments and protects the steel by means of a pure barrier effect.

Coal tar paints possess high water resistance and were widely used for immersed structural details in the offshore industry and in the ballast tanks on ships. Coal tar coatings are not used these days due to the increased risk of eczema, skin, lung, bladder and kidney cancer. They have also been classed as a carcinogen.

Polyesters and Vinyl Esters

The major use of polyester and vinyl ester coating systems is for solvent and chemical resistance purposes. As a result they are commonly grouped under "tank linings" but extensive use is also seen for containment areas and as part of laminate systems for structural design of vessels and structural repair of tanks and boats.

The original use of these resins was in mortars for chemical resistant brick linings and fibre glass reinforced structures. The basic structure of these coatings involves the initial combination of the thermosetting resin pre-polymer (polyester or vinyl ester) with an unsaturated monomer, usually styrene. In the application stage this resin is reacted with a peroxide catalyst, usually methyl ethyl ketone peroxide (MEKP).

The result is a film former that has theoretical volume solids of 100%. However, it is not unusual to have dry film shrinkage of 3% to 15% due to evaporation of the styrene prior to reaction with MEKP under hot, dry and windy conditions. The vinyl esters have broader chemical and solvent resistance than the polyesters due to the fact that these resins are formed from a reaction of epoxy resin with acrylic or methacrylic acid.

Polyesters and vinyl esters normally resist temperatures of 100°C wet and 177°C dry. Usually blast cleaning to Sa2.5/Sa3 is required before such products are applied. The roughness of the steel should be somewhat higher than is normally the case for other paints.

Polyester glassflake paints are usually applied by large high-pressure pumps of 60:1 ratio. Large nozzles are used, often in the range of 0.040–0.050 inches. The pot life is short, often less than one hour at room temperature. Like with all two-component products it is important to clean hoses, pumps and other equipment thoroughly after use. If this is not done the paint will eventually cure and even the strongest solvent will not dissolve it.

Polyurethane Paints

Two component polyurethane paints have several properties in common with epoxy paints. When the components are mixed a chemical reaction takes place and a three-dimensional insoluble network is formed. Polyurethane paints can generally be used at lower temperatures than epoxy paints. Some polyurethanes can be applied at sub-zero temperatures. The polyurethane paints have better weatherability and sunlight resistance than epoxy paints.

For finishing coats an aliphatic polyisocyanate type curing agent is used and tends to yellow less. For primers and intermediate coats aromatic isocyanate type curing agents can be used. Polyurethane paints are sensitive to high relative humidity of the air during application. Note that the isocyanates often react faster with water in the air than with the base resulting in insufficient cure. One-component polyurethanes are produced which require humidity to cure.

Advantages of Two-Component Polyurethanes

- Wide range of application temperatures from below 10°C to 43°C
- Compatibility with a wide range of existing organic coatings
- Extended recoating intervals
- Excellent colour and gloss retention.

Disadvantages of Two Component Polyurethanes

- Sensitive to moisture during curing with loss of gloss and wrinkling
- Less ability to withstand immersion or chemical attack than epoxies or aromatic polyurethanes
- Worker sensitive to free isocyanates during application.

Moisture-Cured Polyurethane

Moisture-cured polyurethane coatings were introduced into the United States in 1955, but the polyisocyanate chemistry dated back to Otto Bayer's work in Germany in the late 1930s. For moisture-cured polyurethanes the predominant reaction of concern is that of the isocyanate reactive group with moisture. When humidity or moisture is high, the formation of carbon dioxide during this reaction can cause bubbling.

When the applied film is too thick, the carbon dioxide can become trapped in the film, also causing bubbling. Moisture-cured polyurethane coatings are also available in both aromatic and aliphatic versions and combinations of both. The advantages and disadvantages of aromatic versus aliphatic versions are similar to those of two-component polyurethanes.

Advantages of Moisture-Cured Polyurethanes

- Adhesion to a variety of substrates is usually excellent because of their wetting ability. Once they penetrate into the substrate they react with available moisture and active hydrogen to create a very tight bond.
- Adhesion to damp marginally cleaned surfaces is better than two component polyurethanes because of the isocyanate/moisture reaction during the curing process.
- Aromatic versions have very good chemical and solvent resistance.
- Good abrasion resistance.
- Good impact resistance.
- Will cure at low temperatures depending on formulation. However, the cure rate is much slower at these low temperatures.

Disadvantages of Moisture-Cured Polyurethanes

- Application on surfaces that have too much moisture can cause bubbling from the formation of carbon dioxide, resulting in pinholes and voids in the paint film.
- Aromatic versions have poor resistance to UV light with loss of gloss and chalking.
- Limited pot life after mixing due to moisture reaction with isocyanate in the mix.
- Ingredients are generally more expensive than two-component polyurethanes.

Polysiloxanes

The development of the most intriguing combination of organic and inorganic film formers has opened a completely new range of possibilities for both new construction and maintenance coating programmes. However, just like the slow commercial development of inorganic zinc primers during the 1940s and 1950s, chemists have had to develop some unique formulation parameters to overcome the compatibility limits of the organic and inorganic components of the coating.

The term "polysiloxanes" can include silicones but is used here in its broadest sense, that is any polymeric structure that contains repeating silicon oxygen groups in the backbone, side chains, segments or crosslinks regardless of the substitution of the silicon atom. The chemistry first commercialized in the United States has now spread to the United Kingdom and Europe.

Several pure polysiloxane network compositions have maximized thermal and chemical resistance pure polysiloxane compositions offer extended weatherability and appearance retention, and hybrid systems in which properties of a traditional resin have been selectively and significantly upgraded.

Heat resistant pure polysiloxane coatings have been formulated that provide heat resistance in excess of 1,100°C. This technology can be used in an essentially inorganic formulation that behaves like a zinc silicate without the acid exposure and chemical reactivity limitations of inorganic zincs. Polysiloxane acrylic top coat blends provide colour and gloss retention equal to or greater than traditional aliphatic polyurethanes.

The most significant commercial use of polysiloxane technology has been the incorporation of the polysiloxane into an epoxy matrix that is inherently compatible with inorganic zinc primers. The corrosion resistant properties of this two-coat system meet or exceed the performance of traditional inorganic zinc/epoxy/polyurethane three coat systems. Early versions of these products were limited to two component epoxy products.

Later versions have included single component products based on synthetic polymer backbones. In many ways this inorganic/organic hybrid polysiloxane chemistry can be considered the most significant advance in the ambient cure corrosion protection industry for many years, meaning the current and likely future requirements for VOC and health and safety along with reduced application costs, and from performance projections, lower life cycle costs due to increased weatherability and durability when compared with polyurethanes.

Another version of polysiloxane chemistry is based on urethane acrylate modified polysiloxane products currently in use in the United Kingdom, Middle East, and Europe. Recent tests in the United Kingdom and Norway have demonstrated that two-coat polysiloxane systems gives at least as good and often better performance than a conventional zinc rich epoxy primer, epoxy intermediate coat, polyurethane finish, totalling 325μm.

The anticipated service lifetime of these new polysiloxane systems range from 13 to 37 years depending on the service environment. The urethane acrylate backbone of these products is said to give better gloss and colour retention, better flexibility and better impact resistance than the epoxy backbone polysiloxane.

Solvent-Free Epoxy

The high-build solvent-free epoxies are becoming more and more popular due to the problems involved with the use of solvents and air pollution. There are a number of materials now on the market referred to as high solids solvent free epoxies. These are made with a sufficient amount of reactive diluent to reduce the viscosity so that they can be properly handled by normal spray units. Actually there are not always 100% solid epoxy coatings since they can contain a very small amount of solvent in addition to the reactive diluent.

Most of these materials react to form a coating according to the amine type reaction and may use the ketamine-blocked amine type of approach. There are a number of advantages to the solvent-free epoxy coatings. Some can be applied by ordinary paint spray equipment particularly the coatings with somewhat less than 100% solids. They build a heavy film and react in place to form a very good corrosion resistant coating.

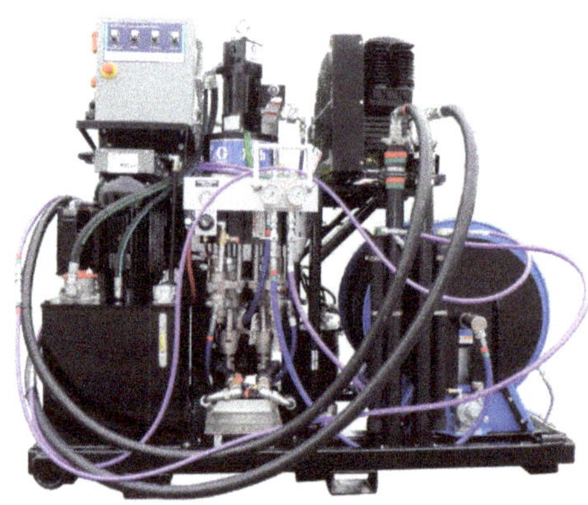

Plural Component Spray Pump

Truly 100% solids epoxy coatings generally use a two-part gun that meters the amine into the system at or near the gun tip. In both cases, although to a greater extent with 100% solids coatings, care should be taken not to build thickness in excess of the paint manufacturer's recommendations. Each coating has its own inherent stress level and over building the coating may cause cracking, particularly at stress points in the steel.

The high solids materials often are also applied as a hot mix coating. If such a hot resin mixture is applied to a cold substrate, poor wetting can result with very poor adhesion of the coating particularly if applied in a heavy film. These problems should be taken into consideration during application where high solids epoxies are involved. Newer epoxy resin modifications have resulted in less thixotropic formulations on 100% solid epoxy coatings with longer reaction times, greater film flexibility and most importantly greater penetrating properties.

Epoxy Phenolics

The phenolic epoxy coating is another instance where the hydroxyl group on the epoxy resin is a primary area of activity. Many of the epoxy phenolic coatings are baked coatings requiring several hundred degrees of temperature to properly cure. There are also epoxy phenolic coatings available which will cure at low temperatures.

They are very slow to cure at ambient temperatures; however, at approximately 60°C they do cure to a chemical resistant coating with very nearly the equivalent resistance to baked coatings. Many of these coatings are used for solvents, saltwater and particularly for the lining of tank cars or barges transporting concentrated sodium hydroxide solution. These epoxy phenolic coatings or linings provide excellent corrosion and chemical resistance.

Glass Flake Epoxy

Glass flake epoxy coatings are used to protect steel in extremely aggressive environments. When these coatings were first introduced, they were primarily used in offshore applications. In recent years, however, they have been gaining acceptance in mainstream infrastructure as well. Glass flake pigments are wide and thin, and when the paint flows out onto the surface during application, they tend to orient themselves parallel to the substrate.

This allows them to form layers with a large degree of overlap, creating a highly effective barrier against moisture and chemical penetration because the pathway around and between the glass flakes is extremely long. The glass pigment can also confer increased impact and abrasion resistance and may aid in relieving internal stress of the cured coating.

Epoxy Curing Agents

The performance properties of the two or three component epoxy coatings are further enhanced by the choice of curing agents. The more common ones and the properties they bring to the particular epoxy are the following:

Aliphatic Polyamine: These are multi-functional, meaning they react on more than one site on the molecule of resin. The result is a highly crosslinked, tough film with high acid and alkali resistance. They generally have short pot life and curing times. They also cause dermititis in workers from direct skin contact. One disadvantage is their tendency to react with carbon dioxide and moisture, particularly at low temperatures, to form an amber coloured, sticky, water soluble film on the outside of the coating known as "amine blush" which must be removed before over-coating.

Polyamine Adducts: These are made by partially reacting an aliphatic amine with epoxy resins to form amine adducts. This so called "kick starting" of the epoxy curing reaction tends to result in longer pot life, less reactive mixing ratios, and less tendency to amine blush. Another advantage is the higher viscosity, which helps film building properties.

Ketamines: These are aliphatic amines reacted with ketones to produce "blocked" amines meaning that the amine cannot cross link (cure) until it is unblocked by reaction with available moisture in the atmosphere. Curing is relatively slow and excessively thick films may never cure thoroughly. Properly cured, they have very good acid and alkali resistance and are often used in high build tank linings.

Polyamides/Amido-amines: These are very user friendly with good pot life and a wide range of mixing ratios. They tend to be more flexible and resistant to moisture, weather and alkali, but have poor chalk resistance, and only moderate acid and alkali resistance. They normally require an induction period after mixing prior to application.

Water and corrosion resistance are improved over straight amine systems. They can generally be steam cleaned or detergent washed repeatedly without deterioration. Cure proceeds very slow below 10°C. The amido-amines are excellent wetters and can displace water from damp surfaces, particularly when the coating is brushed or rolled.

Mannich Bases: These amines are modified with methylol phenol in an accelerating process. They are valuable as curing agents for blush resistant epoxy systems that cure at temperatures of -1 to 2. Their resistance to chemicals is generally better than polyamides, but less than aromatic amines.

Aromatic Amines: Aromatic is a chemical term meaning an unsaturated ring of carbon atoms. Although giving excellent chemical resistance, the more popular of the amines in this group is a suspected animal carcinogen, and amines used to replace it do not generally have the same chemical resistance properties. All aromatic amines chalk easily.

Cycloaliphatic Amines: The performance properties of these amines are somewhat between those of aliphatic/aromatic amines and polyamides. They have very good surface wetting properties and produce a tight, tough barrier film, making them a good choice for surface tolerant epoxies. They give high gloss, non-blushing films and often do not require an induction period.

Polyisocyanates: These are simply a reaction between the isocyanate resins, such as those used to cure polyurethanes, and the epoxy resin. The film cures fast with good flexibility and barrier properties, but is somewhat sensitive to moisture during the curing process. They exhibit very good low temperature curing characteristics.

Phenalkamines: These curing agents have good low temperature curing properties and require little or no induction time. Wetting and adhesion properties are excellent, so they are popular as curing agents for "surface tolerant epoxy products". They have good acid and alkali resistance as well as good fresh and saltwater resistance. Early versions did not have good UV resistance. Recent versions have improved resistance to UV attack and chalking.

CHAPTER EIGHT
Paint Application Equipment

They are many methods used to apply a protective coating for industrial use; these include:

- Brush
- Roller
- Airless Spray
- Conventional Spray
- Electrostatic Spraying
- Plural Component Spraying.

Paint Brush

Application by brush is the traditional method that has been used for centuries, even used by the ancient Egyptians. Brush application is particularly suited for working the paint into the substrate or "wetting" the substrate, and is ideal for difficult areas and manual welds; however, brush application has largely been superseded by spray application. Brush application is one of the slowest methods of paint application and results in a thinner and more uneven layer of paint with no more than 50μm dry film applied per coat.

Brush application is also ideal for the application of coatings on surfaces that cannot be properly cleaned or prepared. For surfaces that are deeply pitted the superior wetting action of the brush can penetrate deep down into the pits whereas spray application will leave the paint sitting on top of the pits unable to penetrate, therefore providing little or no adhesion.

When applying paint to small areas for maintenance it may be useful to apply the first coat by brush if possible due to the wetting action; however, the paint manufacturer's recommendations should be verified. When applying paint by brush it is dipped into the can and the excess material is removed by wiping the brush on the side of the can as it is withdrawn.

A tapping action of the brush may also be useful to remove excess paint. Most of the anti-corrosive paints have poor brushability properties – this is because they have short pot lives and cure before they have been completely smoothed or brushed out, leading to a defect known as "ropiness".

Flat Paint Brush

Brush application should be made at right angles in a cross-hatch style as this procedure ensures a complete and even coverage and helps prevent defects such as pinholes or bare spots. Do not brush against the edges as the coating may be pulled from the surface, always brush onto an edge as this will ensure good coating thickness. It important not to press down on the brush during application as even strokes are required.

Advantage of Brush Application

- Easy to use
- Good for stripe coating
- Easy to clean
- No power source required
- Superior wetting action
- Relatively inexpensive.

Disadvantages of Brush Application

- Results in thin coats
- High builds are difficult to achieve
- Produces uneven layers

- Slow compared to other methods
- Labour intensive.

Particular attention must be given to edges as these will be one of the areas first to break down; this is known as "edge failure". When edges are coated, the coating has a tendency to shrink and pull back during the curing process as the coating flows away – this is known as "surface tension". Stripe coating edges is a common method for ensuring that enough paint is applied, and brush application is the preferred method as the brush has superior "wetting" properties and works the coating into the substrate.

The brush can fill in any "surface irregularities" like porosity and small voids which may be present on the substrate. Even minute particles of dust may be brushed into the surface, although this is not recommended. Edges are normally radiused to 2mm as this will help eliminate the coating shrinking and pulling back. If the edge is "sharp" it could lead to premature coating failure.

Roller

Roller application is ideal for use on large, flat areas. It is quicker than brush application but not as fast as airless or conventional spray. Roller application is a popular method of application in the painting and decorating industry to apply emulsion to walls and ceilings. It is also used extensively in the protective coatings industry; however, not all coatings can be applied by roller, as many high performance coatings need to be applied by airless spray. Using a roller is quite straightforward to use. The paint is stored in a tray with a perforated grid and the roller dipped into the paint then rolled over the grid to remove excess material.

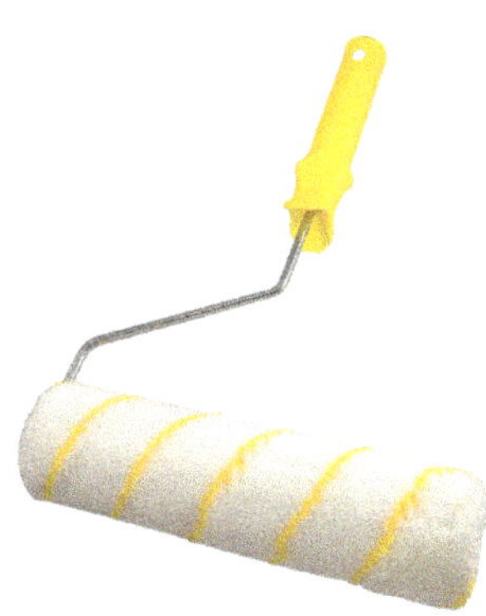

Roller

Within the tray a reservoir is at one end and holds the paint, and after loading the roller it is rolled out on the platform of the tray. When we use rollers the paint is applied as a "W" and rolling must start vertically and erase the "W" horizontally. The roller will not achieve high film builds, and is not compatible with certain coatings especially products with a short pot life or heavily pigmented coatings.

The quality of the film produced is also dependent on the "nap" of the roller. This term refers to the make-up of the fabric covering (length and density) and effects the quantity of paint applied and the texture of the applied film. Use a thick nap on pitted or uneven surfaces and a thin nap where appearances are important. Whichever nap is used, a textured or mottled effect is usually found with roller application.

Also, when using the roller, air may be pushed into the coating and the trapped air will produce small blisters on the surface of the paint. It may be necessary to "lay off" the rollered surface with a brush by passing it vertically and horizontally over the surface for a smoother and more aesthetic finish. Selection of the correct length of roller nap or pile for the particular coating is critical to successful application. The advice of the coating manufacturer should be sought to determine the best type of roller material to be used.

Various types of roller materials are available and these include:

- Nylon
- Lambswool
- Mohair
- Plastic Foam.

Roller Heads

Paint rollers are produced in lengths of 13cm, 18cm and 25cm. The quality of the roller can vary tremendously between suppliers and associated equipment. The roller head or cover slides over the roller mechanism and can be left and cleaned in-situ until a new roller head is required. Washing the new roller head in freshwater will prevent the roller fabric material from coming loose and contaminating the freshly coated surface. Some rollers do not require a pre-clean and can be used as delivered. The roller head can last quite a long time as long as it is not abused and incorrectly stored and therefore must be cleaned immediately after use.

Advantages of Roller Application

- Roller application is faster than brush
- Very easy to use
- The roller head is available in a number of fabrics
- No power source required.

Disadvantages of Roller Application

- Slower than airless and conventional spray
- Labour intensive
- Roller application will not penetrate pitted or rough surfaces
- Roller application may result in low film thickness
- No good for high solids type coatings.

Airless Spraying

Airless spraying does not use compressed air to atomise the paint; instead the paint is forced through the nozzle orifice at high pressure by piston pressure. As the paint leaves the nozzle orifice it expands into a "spray fan" which then allows the applicator to apply a wet film to the surface being treated. This spray fan is controlled by the tip size, angle and pressure used during application. In airless spray equipment the material is under high pressure between the paint pump and the spray gun. As air is not used to atomise the paint hence the term "airless"; however, unlike conventional spraying the material is not under pressure in the container.

The Paint Pump consists of two main parts:

- Air Motor
- Hydraulic Cylinder

The main component of the **air motor** is the **air piston**. The valve mechanism governs the compressed air so that is acts alternately on the upper part and on the lower part of the piston. The air motor is coupled to the hydraulic cylinder by a transmission rod. **The hydraulic cylinder** consists of a lower cylinder chamber, a piston, a piston rod, an upper cylinder chamber, an upper and lower ball valve and upper and lower packing seats.

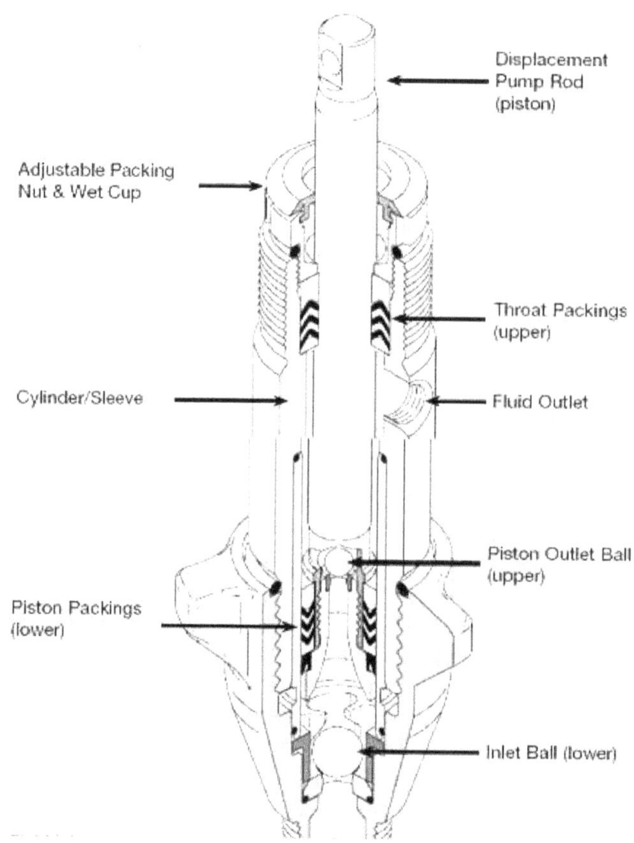

Pressure is produced in the lower part of the cylinder when the piston rod moves upwards. When the inlet valve opens, paint is siphoned into the chamber below the piston, and at the same time the ball valve in the piston closes. The paint in the upper part of the cylinder is forced out into the hose. When the piston moves downwards the paint presses against the inlet valve and closes it, at the same time the paint presses against the ball valve in the piston which opens.

The paint passes through the open ball valve and flows into the upper part of the cylinder and further out into the hose. The pump applies full pressure to the paint during the upward as well as downward strokes. The relationship between the piston area in the air motor and the piston area in the hydraulic cylinder indicates the pump ratio. If a pump has a ratio of 10:1 it means that the surface area of the air motor piston is 10 times greater than the paint piston surface area.

If you put 6.8 bar (100psi) air pressure onto the air motor the resultant paint pressure will be 10 x greater, i.e. 6.8 bar (100 psi) air x 10 = 68 bar (1000 psi) paint pressure. When the spray gun is deactivated the paint pump will continue working until the air pressure and the paint pressure balance. When the gun is reactivated the paint pressure decreases; at this point the pump will start to maintain the paint pressure by replacing the volume flowing out of the nozzle orifice.

Advantages of Airless Spray

One major advantage is the high rate of fluid delivery which permits greater production rates.

- Airless spraying is especially appropriate for applying highly viscous protective coating to large exterior surfaces such as bridges or storage tanks.
- Atomization is established by the tip selection thus minimizing opportunities for operators to mis-adjust the units.
- Airless guns can produce good quality finishes if pressures are kept low in applications where special tips are used.
- Lower viscosity materials spray quite well at these lower pressures.
- Airless spray coating technology causes less overspray or bounce back than airspray guns and because of the high velocity of fluid emanating from the tip during operation, airless spray guns achieve good penetration on recessed areas of work pieces.

Disadvantages of Airless Spray

- A serious concern with airless spray technology is the potential risk of injection injury.
- Another issue is that the operator has no control over the amount of fluid flow, the trigger activates an all-or-none fluid flow (no feathering as is possible with

airspray guns). This requires greater operator skill on smaller parts.

- Airless spray coating technology is generally not considered capable of providing a very high finish quality. The initial investment with this technology is relatively high due to the need for a high pressure pump.
- A final limitation is increased part wear since the necessary high pressures cause abrasive fluids to wear out key parts very rapidly.

The following equipment is used with airless spraying

- Paint line or fluid hose
- Airless spray tips
- Reversible tips
- Airless spray gun
- Fine filters
- Swivel coupling.

Paint Line (Fluid Hose)

The paint line or fluid hose must be able to withstand the load of the high pressures supplied by the pump in all conditions. They must be made from solvent resistant materials and be able to tolerate high ductile and bending loads. The paint hose material will consist of an inner tube, metal carcass and outer protective jacket. The paint line requires re-enforcement due to the high pressures under which it is used as one of the biggest risks to the operator of spray equipment is injection of paint material directly in the body.

Airless Spray Gun with Fluid Line

For health and safety reasons an earthing wire is also braided in which must be connected to the gun at one end and to the airless pump at the other and from there to electric earth; the earthing is to divert charged static electricity. If the spray unit is not earthed electric sparks may be discharged from the gun during paint application. If paint containing flammable solvents is sprayed in tanks or poorly ventilated areas the discharge may ignite the critical explosive mixture of solvent and air.

The paint lines must have the correct dimensions in order to carry enough paint to the spray gun without too much loss of pressure. In the case of short hose lengths and light fluid materials, hoses with ¼ inch internal diameter can be used. The most widely used paint lines for spraying of anti-corrosive coatings have an internal diameter of ⅜ of an inch; they provide flexibility because it is possible to transport most types of coatings over varying distances without loss of pressure. The thickness of the ⅜ of an inch paint lines makes them rigid to handle; normally a ¼ of an inch whip end is inserted between the ⅜ of an inch line and the gun, then the gun can be handled more freely.

Airless Spray tips

Coating fluids under high pressure can be highly abrasive. Key parts of airless spray guns such as the fluid valve, fluid needle, seat and the spray tip are made from tungsten carbide or other hard materials to provide an acceptable service life. Spray tips for airless spray coating have a unique design, causing them to form a fan-shaped spray with high pressure fluid. The tip orifice is formed by cutting a V-shaped groove in the front end of the tip deeply enough to extend into the internal cavity of the tip.

Airless Spray Tip

Airless spray tips are not adjustable. Each nozzle produces a spray with a characteristic fan pattern width and flow capacity. Spray gun suppliers offer a range of tips with different fan sizes and flow capacities to suit a wide range of applications. While the spray fan is sometimes stated in degrees from 5° to 95° it is more often stated in terms of fan pattern width in inches at some distance from the substrate. For spraying of flat surfaces, tips ranging from 60° to 80° are used. Flow capacity is usually stated in terms of gallons or litres per minute. The orifice size is generally stated in thousandths of an inch or millimetres.

Tip Selection: The user typically will select a tip with the desired fan size and then experiment with different orifice sizes to find the one that best suits the job. Painting contractors normally have at their disposal a selection of spray tips from 0.015° to 0.027°. Typically there will be tails in the spray pattern if the orifice size is too large, the flow rate will be too high or the pressure will be too low to form a good spray pattern. Conversely if the orifice is too small the fluid flow will be too low to maintain acceptable productivity.

Reversible Tips

By far the most popular tip designed for airless spray is the reversible tip as the tip can be cleaned without any adjustment or removal from the airless spray gun. A general rule of thumb with spray tip wear is that if the fan size has decreased more than 25% it is time to replace the tip.

Extending the Life of a Spray Tip

- Use the correct spray tip
- Strain the paint and/or use filters in the equipment
- Keep the pressure to a minimum
- Look after the tip by cleaning with a soft brush and correct solvent.

General Rules on Tip Selection

- Small orifice size is used for thin material
- Large orifice size is used for thick material
- Greater fluid flow is found with large orifice
- Faster coverage is found with large orifice.

Airless Spray Gun

Fluid is pumped to the gun under high pressure. It is transmitted by the fluid hose and enters the gun via the fluid inlet and flows to the fluid valve. The fluid valve controls the fluid flow. When the valve is closed fluid cannot flow any further; when the valve is open fluid flows under pressure to the spray tip. To operate the gun the user pulls the trigger toward the handle which opens the fluid valve.

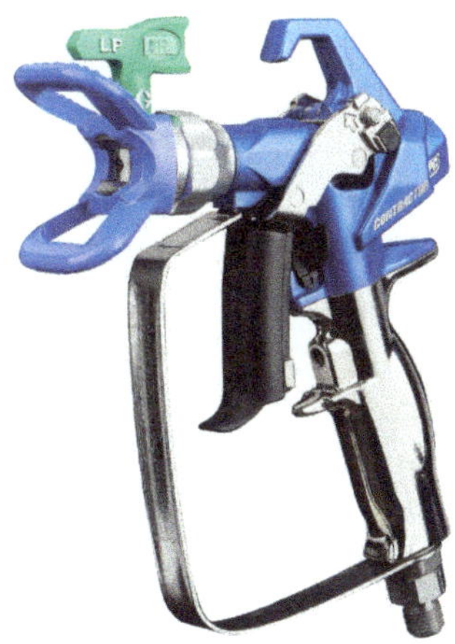

Airless Spray Gun

Fluid travels past the diffuser to the spray tip via internal passages in the gun body. Fluid flows out the spray tip at high velocities as fluid is released from the orifice; it atomizes into small droplets and travels to the substrate under its own initial momentum. When the operator releases the trigger, springs close the fluid valve, stopping the fluid flow.

Since the high pressures of airless spray guns have the potential for causing skin injection injury, it is important to know about and comply with all the necessary safety precautions. In addition to avoiding skin injection injury, operators must ground the fluid hose and fluid supply to prevent static electricity build-up since static electricity can cause fire, explosion or electric shock.

Principal Parts of an Airless Spray Gun

The principal parts of an airless spray gun include:

- **Spray Tip:** The spray tip controls the fluid flow rate and creates back pressure in the system. The spray tip also defines the spray pattern by the size and shape of the orifice. Typically a small orifice is used for low viscosity (lighter) fluids and a larger orifice for higher viscosity (heavier) fluids. The fluid tip can be rotated to any position for ease of operation.
- **Trigger:** Squeezing the trigger activates the fluid valve which permits fluid to flow to the spray tip. Airless guns do not have a variable valve as do air spray guns so "feathering" is not possible.
- **Tip Guard – (Duck's Bill):** The tip guard consists of two protrusions on either side of the spray tip. The tip guard warns the operator to stay clear of the tip to prevent skin injection injuries.
- **Trigger Lock:** The trigger lock is a safety feature that prevents the trigger from activating accidentally. The trigger lock should be engaged any time the gun is not in use.
- **Diffuser:** Helps the efficiency of atomization. The diffuser has a 2.28mm orifice with a bar inside which splits the high pressure stream of material. The diffuser is also a built-in safety device, designed to break up the high pressure coating stream should the gun be triggered without a spray tip in position to atomize it.
- **Fluid Inlet:** The fluid inlet connector serves as the attachment for the high pressure fluid supply hose.
- **Fluid Valve:** The fluid valve consists of the fluid needle and seat. This valve controls the flow of fluid to the spray tip with an off/on ball seat valve and is mechanically actuated by squeezing the trigger. The needle connects the trigger to the ball.
- **Gun Body:** The gun body is an ergonomically-designed aluminium or plastic housing that serves as a frame for internal parts, a container for fluid passages and as a handle for the operator.

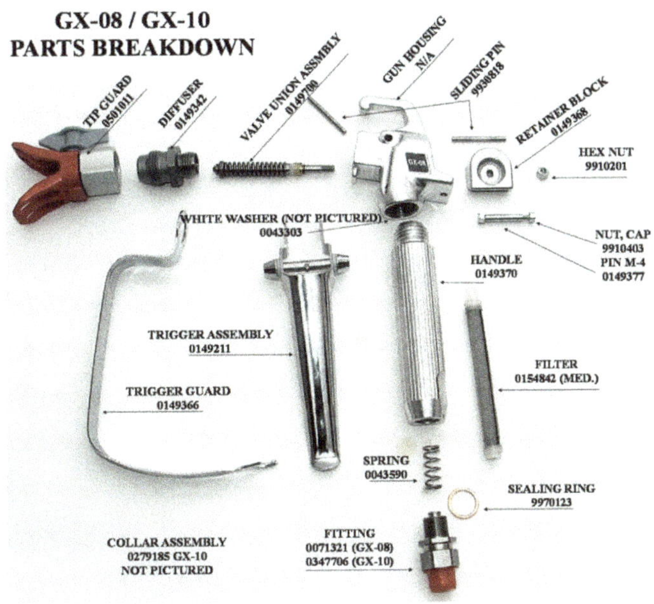

Airless Spray Gun Parts Breakdown

Fine Filter

Even if the pump has a coarse filter and both pump and paint hose are cleaned on a regular basis, contaminants may reach the spray tip and block it up. These contaminants are so small that they may slip through the coarse filter. This type of contamination must be strained from the paint on its way from the paint outlet on the pump to the paint tip – for this purpose fine filters are used. The filters can be mounted as a separate unit after the paint outlet, in the pulse chamber, in the swivel coupling or in the spray gun.

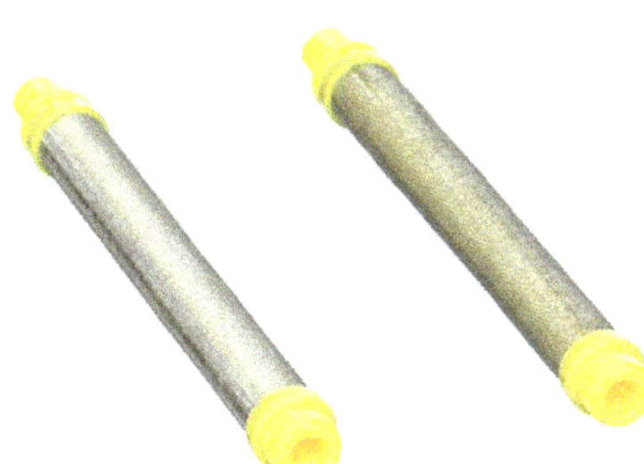

Fine Filters

The smaller the nozzle orifice the smaller the particle it takes to block it. Consequently, the nozzle orifice determines how fine the filters can be. The nozzle orifice is usually given in thousandths of an inch with the filters in mesh size. The mesh size is the number of meshes per inch.

- Nozzles with an orifice of 0.015" or less require a 200 mesh filter
- Nozzles with an orifice between 0.15-0.021" require a 100 mesh filter
- Nozzles with an orifice between 0.21-0.031" require a 60 mesh filter.

Swivel Coupling

The swivel coupling is also a facility for obtaining easy handling of the gun. It has the effect that the gun can be turned round freely without the paint line following the gun around. The swivel coupling is inserted between the gun and the whip hose.

Maintenance of The Spray Equipment

Maintenance of the airless spray equipment mainly consists of oiling and replacing worn parts, and should be overhauled by skilled and trained technicians. Fluid hoses and couplings must be checked and replaced when damaged or broken and when there are any doubts about the quality or strength of the fluid hose materials. Wearing parts in the gun, needle, seat and nozzles are replaced when wear or damage is found. The trigger mechanism will wear with time and must be oiled along with the trigger spring.

Airless Spray Safety

Since the high pressures of airless spray guns have the potential for causing skin injection injury, it is important to know about and comply with all the necessary safety precautions. In addition to avoiding skin injection injury, operators must ground the fluid hose and fluid supply to prevent static electricity build-up since static electricity can cause fire, explosion, or electric shock.

- A pressurized unit should never be left unattended. The unit should be shut off, the pressure relieved, the spray gun's trigger safety engaged, and the power shut off before leaving.
- All fluid connections should be high-pressure rated airless spray fittings,

tightened securely and checked before each use.

- The fluid hose should be electrically grounded to reduce the hazard of static electricity sparking.
- The coating and solvent manufacturers' safety precautions and warnings should be followed.
- Any unsafe condition or practice should be reported to the safety supervisor immediately.

Conventional Spraying

With conventional spraying the coating is atomized by a stream of compressed air and is carried to the surface on a current of air. Both air and coating enter the gun through separate passages (channels), are mixed and are driven through the air cap in a controlled spray pattern.

Conventional Spray Guns are Divided into Two Types:

- Guns provided with a cup mounted directly on top or underneath the gun – these are referred to as gravity guns and suction cup guns.
- Guns attached to a separate pressurized tank – these are referred to as pressure feed guns.

These Two Types Can Be Divided into:

- Guns with internal or external mixing of atomizing air and paint
- Guns with suction feed and pressure feed.

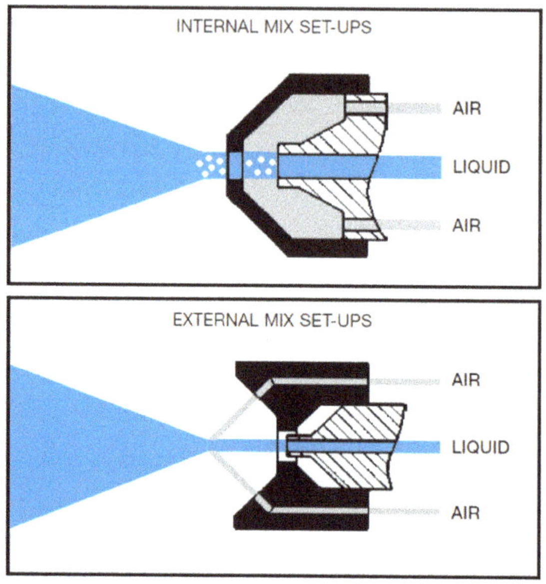

Internal Mix Gun

This gun mixes air and material inside the air cap, before expelling them. It is usually used where low air pressures and volumes are available, or where slow-drying materials are being sprayed. A typical example is spraying flat wall paint or outside house paint, with a small compressor. Internal mix guns are rarely used for finishing when a fast-drying material is being sprayed, or when a high quality finish is required.

Internal-Mix Advantages

- Requires less air volume (L/min, cfm)
- Requires less air pressure (MPa, psi)
- Provides minimum overspray
- Yields maximum film build per pass
- Yields maximum size spray pattern.

Internal-Mix Disadvantages

- Provides coarse atomization
- Spray pattern size is fixed

- Air cap and fluid tip will wear
- Only pressure feed is available.

External Mix Gun

This gun mixes and atomizes air and fluid outside the air cap. It can be used for applying all types of materials, and it is particularly desirable when spraying fast drying paints such as lacquer. It is also used when a higher quality finish is desired.

External-Mix Advantages

- Produces fine atomization
- Provides control of spray pattern size
- Saves on air cap, no wear
- Available with suction and pressure feed.

External-Mix Disadvantages

- Requires more air volume (L/min, cfm)
- Requires high air pressure (MPa, psi)
- Can produce maximum overspray
- Yields minimum film build per pass
- Yields minimum size spray pattern.

Gravity Feed Guns

Gravity is used in the design and operation of this gun. The gravity feed cup is mounted above the centre line of the gun and gravity enhances the fluid flow. If you activate the trigger without the air the fluid in the cup would flow – this is due to gravity. The capacity of the cup is generally 600–650cc. The process is generally used for small application processes and not capable of large volume works – **Gravity feed guns are rarely encountered in industrial coating operations.**

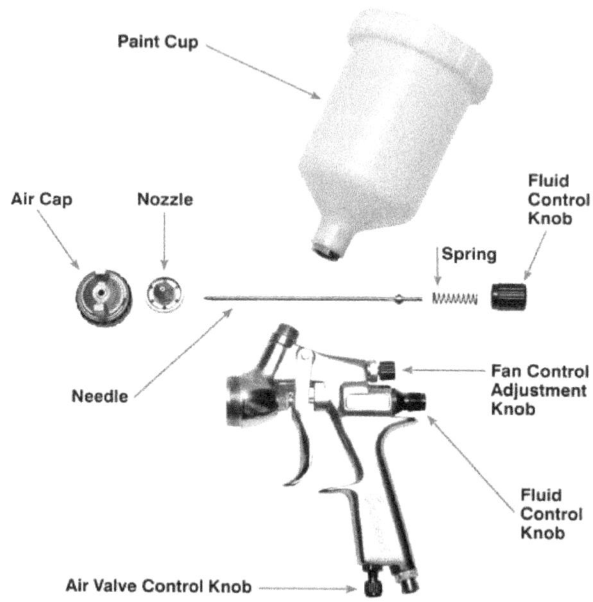

Gravity Feed Gun

Suction Feed Guns

Suction feed guns have a cup mounted below the front of the gun. Suction is created by air flowing from the air cup, which draws the paint from the cup through the fluid nozzle. There are many suppliers of such equipment. The compressed air creates a low pressure area at the air cap, providing a siphoning action. Atmospheric pressure on the paint in the cup forces the paint up a suction tube, into the gun itself and out of the fluid tip. This process of application is ideal for certain small works where colour change is necessary. The equipment is easy to clean and maintain. **Suction feed guns are rarely encountered in industrial coating operations.**

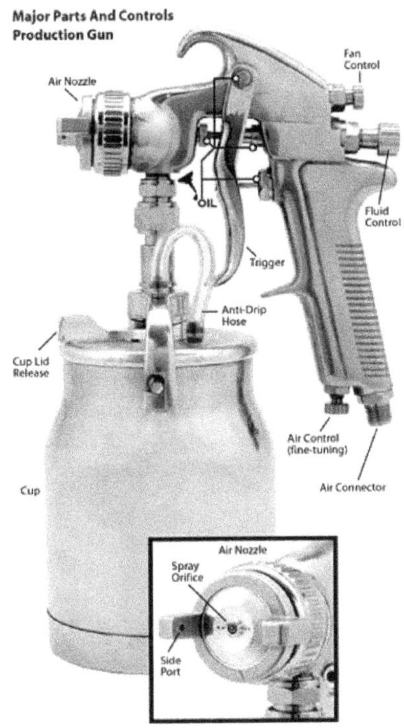

Suction Feed Gun

Pressure Feed System

A Pressure Feed System consists of:

- **Conventional air feed spray gun** – which mixes air and coating and applies the mixture to the work piece in a controlled spray pattern.
- **A pressure feed tank** – contains and holds the coating to be applied.
- **An air filter/regulator** – which helps provide a steady supply of clean, dry air.
- **Appropriate length and diameter of air and fluid hoses** – which transports air and coatings between compressor, spray pot, and spray gun.
- **Air compressor** – provides a continuous supply of compressed air to the material container and spray gun.

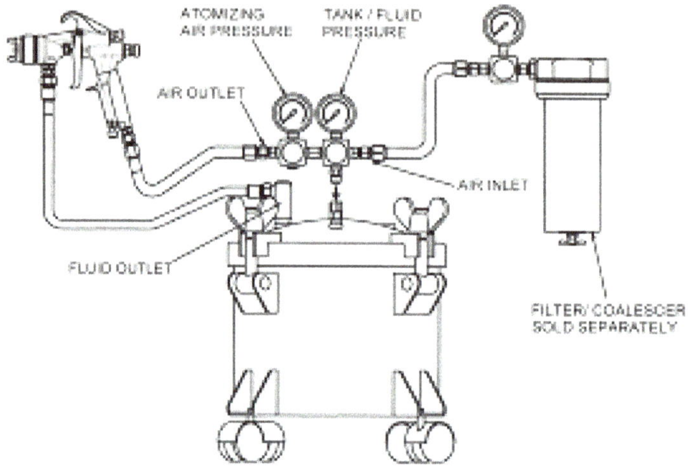

Pressure Feed System

Pressure-Feed Tanks: Pressure-feed tanks are closed containers, ranging in size from about 7.5 L to 450 L. They provide a constant flow of coating material at uniform pressure to the spray gun. The tank is pressurized with compressed air and fluid is forced out of the tank through the fluid hose to the gun. Increasing or decreasing the air pressure in the tank controls the rate of fluid flow.

Pressure-feed tank

Pressure-feed tanks provide a practical, economical method of feeding material to the gun over extended periods of time and are generally used where continuous production is to be maintained. The flow of material is positive and uniform. Relatively high-viscosity and high-solids coatings can be applied in this way, but generally not 100% solids coatings.

Tanks can be equipped with agitators that keep the material mixed and in suspension. There are some modern 100% solids coatings that are very low in viscosity and can be applied by conventional spray.

A Typical Pressure-Feed Tank Consists Of:

- The shell
- Clamp-on lid
- Fluid tube
- Fluid header
- Regulator with gauge
- Safety-relief valve
- Agitator.

The construction of pressure-feed tanks varies according to the service for which they will be used. Light-duty tanks are usually made of welded steel and have inlet air pressure restriction. Heavy-duty tanks are ASME coded and made of drawn steel. When abrasive or corrosive materials are to be sprayed, the tank shell used should be coated or lined with some special material; an insert container can also be used.

Pressure-Feed Tank Regulator

Pressure-feed tanks should be equipped with regulators for three primary reasons:

- Regulators allow material flow to be accurately adjusted until constant.
- Regulators provide adjustment capability to accommodate a wide range of material viscosity and permit pressure increases when added hose lengths demand it.
- Regulators allow more adjustment of the balance between the material flow and the atomization air pressure.

Pressure-feed tanks may be fitted with a single-regulator or a double-regulator. The double-regulator should always be used in industrial spraying applications. One regulator controls the air pressure on the material in the tank, thereby controlling fluid flow. The other controls atomization air pressure to the spray gun. Both controls are at the tank near the operator and can be adjusted for any operating condition. The fluid pressure may be higher or lower than the atomization air pressure depending on:

- Height at which the gun is operated above the tank
- Degree of atomization required
- Size and length of hose
- Speed of operation.

Inspectors should be aware that the single-pressure regulator and gauge found on many conventional spray pressure pots controls only the pressure within the container and not the pressure of the air that atomizes the coating. With a single-regulator the atomizing pressure will be equal to that at the receiver of the compressor (minus the pressure drop in the hose lines); if the receiver pressure cannot be controlled to suit spraying operations, an additional pressure regulator must be inserted in the line between receiver and spray pot or between gun and spray pot.

Conventional Air Spray Gun

The components of a typical manual air spray gun fall into three main categories:

- Control of air flow
- Control of fluid flow
- Gun body.

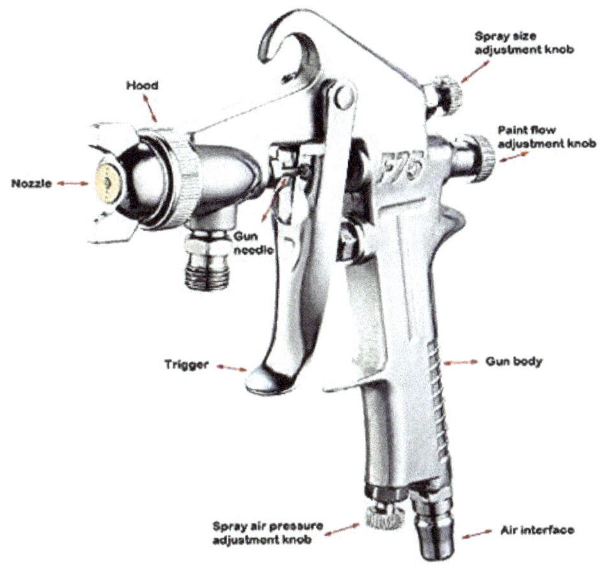

Conventional Air Spray Gun

Air Flow

Some of the major air flow parts are:

- Air-control valve
- Air-adjusting valve
- Pattern-adjusting valve
- Air cap.

Air-Control Valve – The air-control valve is located in the gun's handle, directly behind the trigger. When the trigger is pulled, the air-control valve opens, allowing air to pass through the gun. A positive return spring in the stem of the air-control valve keeps the valve closed until the trigger is pulled. The air-control valve provides no air pressure regulation.

When the air valve is closed, no air enters the gun. When the valve is opened, the full air pressure is admitted. The air pressure is entirely controlled by a pressure regulator located between the compressor and the gun. The air pressure should be adjusted according to the viscosity of the material being sprayed and the desired density of the spray (light or heavy coverage).

Air-Adjusting Valve – An air-adjusting valve (or screw) is provided on some low- and medium-volume guns and is included as an option on high-volume guns. When the valve is built into the gun, it is located adjacent to the air inlet at the base of the handle. When included as an option, the valve is placed at the air inlet connection, in line with the air hose. The air-adjusting valve controls only the flow of air (L/min, cfm), and has no effect on air pressure (MPa, psi).

Pattern-Adjusting Valve – The pattern-adjusting valve (or screw) is located at the back of the gun. It is the top of two knurled knobs found there. The stem of the valve seats in the front of the gun, in either an air baffle or air port. The pattern-adjusting valve regulates air supply to the horns of the air cap. The horn air controls the shape of the spray pattern. When the valve is closed (turned completely clockwise) a round spray pattern results. As the valve is opened (turned counterclockwise) an increasingly narrow fan pattern develops.

Air Cap

The air cap directs compressed air into the fluid stream to atomize it and form the spray pattern.

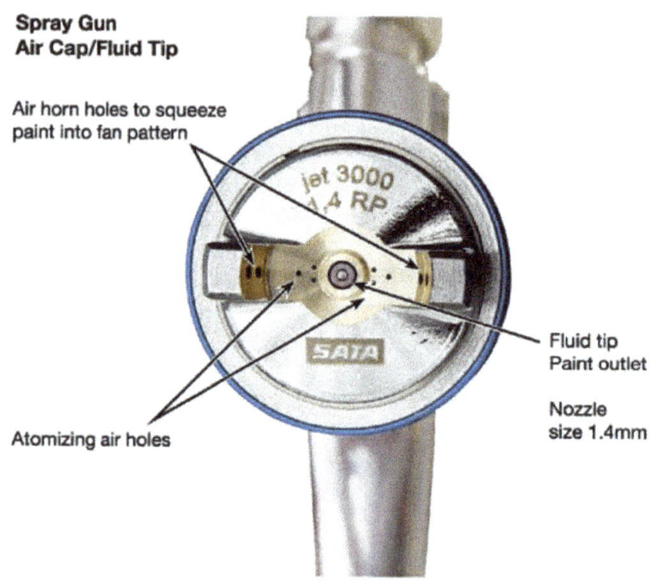

Air Cap/Fluid Tip

Selecting The Air Cap

- Type, viscosity and volume of material to be sprayed
- Size and nature of object, or surface, to be sprayed (multiple or larger orifices increase ability to atomize more material for faster painting of large objects. Fewer, or smaller, orifices usually require less air, produce smaller spray patterns and deliver less material). These caps are designed for painting smaller objects and/or using slower speeds.
- Material feed system used – pressure, siphon or gravity.
- Size of fluid tip to be used (most air caps work best with certain fluid tip/needle combinations).
- Volume of air in cubic feet per minute (CFM) and pressure in pounds per square inch (PSI) available.

Cleaning The Air Cap – The air cap should be cleaned by immersing it in clean solvent and drying it by blowing it off with compressed air. If small holes become clogged, soak the cap in solvent. If reaming is necessary, use a matchstick, broom straw, or any other soft implement. Digging out holes with wire or a nail may permanently damage the cap.

Pulling The Trigger – Operates the spray gun in two stages. Initial trigger movement opens the air valve, allowing atomizing air to flow through the gun. Further movement of the trigger opens the fluid needle, allowing fluid material to flow. When the trigger is released the fluid flow stops before the atomizing air flow. This lead/lag time in the trigger operation assures a full spray pattern when the fluid flow starts. It also assures a full pattern until the fluid flow stops so there is no coarse atomization.

Adjusting The Pressure Feed – Open spreader adjustment valve for maximum pattern size. Open fluid adjustment screw (counterclockwise) until maximum.

Balancing The Pressure Feed Gun For Spraying

- Using control knob on fluid regulator set fluid pressure at 5 to 10 ps.
- Using control knob on air regulator set air atomization pressure at 30–35 psi.
- Spray a test pattern (fast pass) on a piece of paper, cardboard, or wood. From that test pattern, determine if the particle size is small enough and uniform throughout the pattern to achieve the required finish quality. If particle size is too large or is giving too much texture in the finish, turn the atomization pressure up in 3 to 5 psi increments until particle size and texture of finish is acceptable.

- Spray a part with these settings. If you are not able to keep up with the production rate required or if the finish is starved for material, increase the fluid pressure (or use a larger capacity fluid tip) with the fluid regulator control knob in 2 to 4 psi increments until required wet coverage is accomplished.
- Remember as you turn up the fluid pressure the particle size will increase. Once the coverage required is obtained it will be necessary to re-adjust the atomization pressure in 3 to 5 psi increments as explained in step 3 to ensure required particle size and finish texture is achieved.
- If using HVLP using an "Air Cap Test Kit" verify that the air cap pressure in not above 10 psi if required by a regulatory agency.

Fluid Tip and Needle – They restrict and direct the flow of material from the gun into the air stream. The fluid tip forms an internal seat for the tapered fluid needle, which reduces the flow of material as it closes. The amount of material that leaves the front of the gun depends upon the viscosity of the material, the material fluid pressure and the size of the fluid tip opening provided when the needle is unseated from the tip. Fluid tips are available in a variety of sizes to properly handle materials of various types, flow rates and viscosities.

Spreader Adjustment Valve – A valve for controlling the air to the horn holes which regulate the spray pattern from maximum width down to a narrow or round pattern.

Fluid Needle Adjustment – This adjustment controls the travel of the fluid needle, which allows more or less material through the fluid tip. With pressure feed systems, the fluid delivery rate should be adjusted by varying the fluid pressure at the pressure pot. Use the fluid adjustment knob for minor and/or temporary flow control. This will extend the life of the fluid needle and tip.

Air Regulator – This is a device for reducing the main line air pressure as it comes from the compressor. Once set it maintains the required air pressure with minimum fluctuations. Regulators are used in lines already equipped with an air filtration device. Air regulators are available in a wide range of CFM and PSI capacities, with and without pressure gauges and in different degrees of sensitivity and accuracy. They have main line air inlets and regulated or non-regulated air outlets.

Hoses and Connections – The various types of hose used to carry compressed air and fluid material to the spray gun are important parts of the system. Improperly selected or maintained hose can create a number of problems.

- Air hose – used to transfer compressed air from the air source to the gun.
- Fluid hose – used in pressure feed systems to transfer the material from its container to the spray gun.

What Size Air Hoses are Recommended:

- The hose from the compressor to the pressure tank should be a minimum of ⅜" – ½" ID.
- The hoses from the pressure tank regulator to the gun should be a minimum of ⅜"– ½" ID.

Holding the Spray Gun

It should be held so the pattern is perpendicular to the surface at all times. Keep the gun tip 12–18 inches (conventional air spray guns) or 8–12 inches (HVLP guns) from the surface being sprayed. The stroke is made with a free arm motion, keeping the gun at a right angle to the surface at all points of the stroke. Triggering should begin just before the edge of the surface to be sprayed is in line with the gun nozzle.

The trigger should be held fully depressed, and the gun moved in one continuous motion, until the other edge of the object is reached. The trigger is then released, shutting off the fluid flow, but the motion is continued for a few inches until it is reversed for the return stroke. When the edge of the sprayed object is reached on the return stroke, the trigger is again fully depressed and the motion continued across the object.

Lap each stroke 50% over the preceding one. Less than 50% overlap will result in streaks on the finished surface. Move the gun at a constant speed while the trigger is pulled, since the material flows at a constant rate. Another technique of triggering is referred to as "feathering". Feathering allows the operator to limit fluid flow by applying only partial trigger travel.

Spraying Sequence and Technique

Difficult areas such as corners and edges should be sprayed first. Aim directly at the area so that half of the spray covers each side of the edge or corner. Hold the gun an inch or two closer than normal, or screw the spreader adjustment control in a few turns. Needle travel should be only partial by utilizing the "feathering" technique. Either technique will reduce the pattern size. If the gun is just held closer, the stroke will have to be faster to compensate for a normal amount of material being applied to smaller areas.

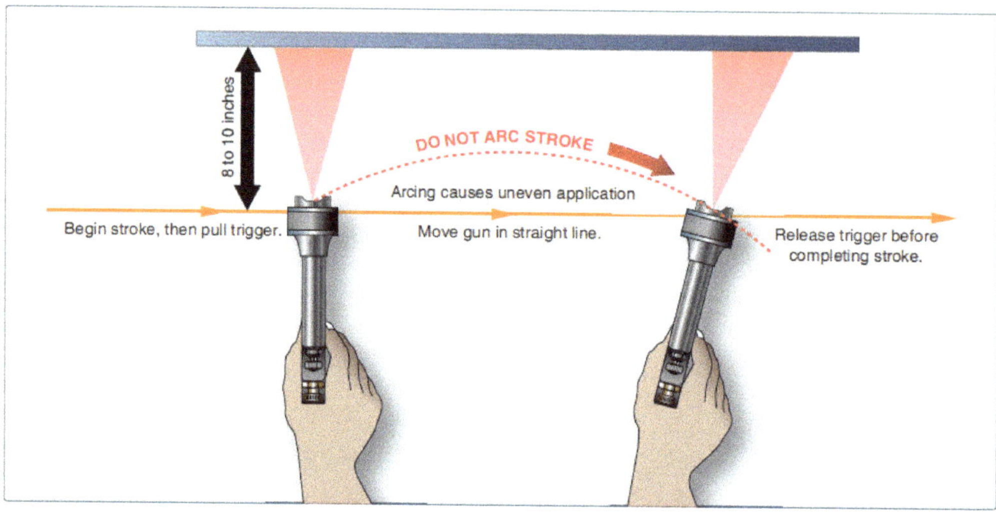

Conventional Spraying Technique

When spraying a curved surface keep the gun at a right angle to that surface and at all times follow the curve. While not always physically possible this is the ideal technique to produce a better, more uniform finish. After the edges, flanges and corners have been sprayed, the flat, or nearly flat, surfaces should be sprayed. Remember to overlap the previously sprayed areas by 50% to avoid streaking.

When painting very narrow surfaces you can switch to a smaller gun, or cap with a smaller spray pattern to avoid readjusting the full size gun. The smaller guns are usually easier to handle in restricted areas. A full size gun could be used, however, by reducing the air pressure and fluid delivery and triggering properly.

Arcing the Spray Gun

Arcing the stroke results in uneven application and excessive overspray at each end of the stroke. When the tip is arced at angle 45°C from the surface approximately 65% of the sprayed material is lost.

Advantages of Conventional Spraying

- Spray pattern is easily adjusted to almost any desired fan width
- High-quality finishes such as for automobiles and furniture can be produced.

Disadvantages of Conventional Spraying

- High loss of coating is caused by overspray
- Billowing and air turbulence are created by the compressed air required
- Reduction of coatings with solvent is often necessary for proper atomization resulting in lowered DFT per application.

Conventional Air Spray Troubleshooting

The spray pattern is very important for the quality of the spray job whether it be airless or conventional spraying. In the protective coatings industry a defective spray pattern is more associated with airless spraying rather than conventional spraying. We will take a closer look at the application faults that may occur during the use of conventional spraying.

Curve To The Left Due To:

- The left "horn" on the aircap is blocked
- The left "supplementary hole" for the atomization is blocked
- Dirt on the outside of the material nozzle (on the left side).

Curve To The Right Due To:

- The right "horn" on the aircap is blocked
- The right supplementary hole for the atomization is blocked
- Dirt on the outside of the material nozzle (on the right side).

Heavy Pattern On Top:

- Dirt on the outside of the material nozzle
- Dirt on the inside of the material nozzle (the underside).

Heavy Pattern On The Bottom:

- Dirt on the outside of the material nozzle
- Dirt on the inside of the material nozzle (the upper side).

Single Split Patterns:

- The atomization pressure has not been adjusted to the material pressure (too high air pressure compared to the material pressure).

Heavy Centre Pattern:

- Too narrow adjustment of the spreader control valve
- Too low atomization pressure
- Too high viscosity
- Too high material pressure compared with the capacity of the material nozzle.

Gun Spit:

- Too high viscosity
- The air hole at the top of the cup is partly clogged
- The connecting nut at the fluid inlet of the gun is loose or damaged
- Loose or damaged nozzle
- The fluid needle seal is in poor condition.

Leakage From The Fluid Needle:

- Loose seal unit
- Worn or dry fluid needle seal.

Dripping From Fluid Tip:

- Dry packing
- Sluggish needle
- Tight packing nut
- Spray head misaligned on type – MBC guns, causing needle to bind.

Excessive Overspray:

- Too much atomization air pressure
- Gun too far from surface
- Improper stroking, i.e., arcing, moving too fast.

Excessive Fog:

- Too much or quick drying thinner
- Too much atomization air pressure.

Fluttering Spray Pattern:

- Lack of sufficient material in container
- Tipping container at excessive angle
- Obstructed fluid passageway
- High-viscosity coatings, requiring addition of thinner
- Loose or cracked fluid pick-up tube in cup or tank
- Loose fluid tip or damaged tip seat.

Dry Spray:

- Air pressure is too high
- Gun is too far from work or out of adjustment.

High Volume Low Pressure - HVLP

HVLP or High Volume/Low Pressure uses a high volume of air (typically between 15–26 CFM) delivered at low pressure (10 PSI or less at the air cap) to atomize paint into a soft, low-velocity pattern of particles. In most cases less than 10 psi is needed in order to atomize. Proper setup utilizes no more fluid and air pressure than is needed to produce the required quality and a flow rate that will meet production requirements.

As a result far less material is lost in overspray, bounce-back and blowback than with conventional air spray. This is why HVLP delivers a dramatically higher transfer efficiency (the amount of solids applied as a percent of solids sprayed) than spray systems using a higher atomizing pressure. The HVLP spray gun resembles a standard spray gun in shape and operation. Models that use high inlet pressure (2080 psi) and convert to low pressure internally within the spray gun are called HVLP conversion guns.

Some HVLP models, particularly those using turbines to generate air, bleed air continuously to minimize back-pressure against the air flow of the turbine. The air cap design is similar to that of a standard spray gun, with a variety of air jets directing the atomizing air into the fluid stream, atomizing it as it leaves the tip. HVLP is growing in popularity and new environmental regulations are requiring it for many applications. HVLP can be used with any low-to-medium solids materials that can be atomized by the gun including two-component paints, urethanes, acrylics, epoxies, enamels, lacquers, primers, etc.

Mixing of Paint

With modern day coatings containing "high volume solids", mechanical mixing is essential so that all of the pigment and binder that lies in the bottom of a paint drum is agitated and dispersed evenly. It is equally important to redisperse all of the paint material that lies internally on the sides of the paint drum; this material may have to be "scraped off" manually with the use of a scraper or a "dogs-leg brush".

The speed of the mechanical mixer plays an important part also. This procedure should be performed as slow as possible to avoid any "air entrapment" in the coating which can in turn lead to "pinholes" during application. The coating should display a homogeneous appearance when mixing; however, mixing at full speed can produce a messy workplace with "splashes" of paint contaminating the work site. During the mechanical mixing process the coating shall display a consistent appearance throughout the paint drum as no colour variations/paint separation is acceptable.

Normally when mixing paints it is mandatory to use mechanical agitation above 5 litres. If manual mixing is called for it is easier to separate the base of the material into separate containers. The two containers can be mixed more easily as the paint clinging to the sides of the container and the sediment lying in the bottom can be removed and mixed thoroughly into a smooth paste. Paints which have been standing for long periods of time normally have sediment lying in the bottom of the can.

Mechanical Paint Paddle

Once both containers have had the paint thoroughly mixed with no traces of sediment visible they can then be poured back into the original container and the curing agent added. Once the curing agent has been added to the base, the paint material can be transferred "back and forth" into a separate container to ensure even dispersion. This process is called "boxing" and once poured back and forth several times the coating will be uniform and should be ready for application.

For some high solids (high-viscosity) coatings including zinc-rich primers, the use of a mechanical mixer may be mandatory and required by the specification or by the manufacturer's technical data sheets. In these cases, hand mixing is not recommended and should not be allowed. If partial kits are to be mixed it is important that the materials are stirred before being measured out and that the measuring is done accurately. Only the required amount of paint should be mixed and any mixed material remaining after the job is completed should be properly disposed of as it cannot be returned to the original containers.

Mechanical Mixing of Paint

General Rules of Paint Mixing:

- Fire extinguishers should always be available.
- Coatings should not be stored too high. Thermometers should be installed.
- Paints should be rotated with the oldest stock being used first.
- The paint store should be clean and tidy.
- Clean any liquid spillage immediately.
- Never smoke in or near a paint store.
- Spill kits should be available.

Paint Storage

Paint storage containers play an important role in any painting project and are seen as an essential part of the paint application process. Paints should be rotated with the oldest products used first, ensuring the product is within its shelf life. The heaviest products should be kept at the lower levels to avoid any back injuries. Keep the base and curing agent together if practical. Always maintain a tidy store and ensure there are no trip hazards on the floor, for example extension cables and paint consumables.

Paint Storage Container

Some coatings can be temperature sensitive therefore paint stores should have minimum and maximum thermometers installed. In hot countries like the UAE, air conditioning units are installed inside of the paint container to maintain a constant temperature. Ideally temperatures for paint storage should be in the range of between 10°C and 25°C (refer to paint data sheet).

Always record the temperatures inside of the storage area at least four times per day and keep a daily log. If paints are left in an unprotected store or left for long periods of time, thickening or settlement may occur, and the older the product the more mixing is required. Always check the shelf life of the product before use; this information can be found on the product data sheets.

Drying Properties of Paint

Drying time is an important characteristic of paints. The length of time for a coating to dry dictates how long one must wait for the coating to be handled or recoated. **ASTM 1640 "Standard Test Method for Drying, Curing or Film Formation of Organic Coatings at Room Temperature"** is one of the methods which defines dry time tests. The procedure involves applying a prescribed thickness of coating. The degree of dryness is determined in accordance with standard definitions:

- **Set to Touch** – The film is set to touch when it still shows a tacky condition, but none of it adheres to a finger lightly touched on the film.

- **Dry to Touch** – The film is considered dry when it no longer adheres to a finger and does not rub up appreciably when the finger is lightly rubbed across the surface.
- **Dry Hard** – The operator exerts a maximum downward pressure without twisting of the thumb on the film and lightly polishes the contacted area with a soft cloth. The film is considered dry hard when any mark left by the thumb is completely removed by the polishing.
- **Dry to Handle** – The film is considered dry to handle when there is no loosening, detachment, wrinkling or other evidence of distortion of the film after bearing down on the film with the thumb, exerting the maximum pressure of the arm, at the same time turning the thumb through an angle of 90° in the plane of the film.

Pot Life

Pot life is defined as "the elapsed time within which a coating can be effectively applied after all components of the coating have been thoroughly mixed". Pot life is a critical issue when applying catalysed coatings. It is important to note that some coatings which have exceeded their pot life may appear to be effectively applied but may not cure properly and may not have adequate adhesion or have poor appearance due to insufficient levelling.

Pot life can be determined by monitoring the change in liquid coating properties over time subsequent to mixing the coating. This may include monitoring viscosity with time or applying the coating to substrates and determining any of a variety of application properties or performance characteristics (sag resistance, levelling, adhesion, gloss).

Induction Period

Induction period is the time interval needed to start a reaction or further reaction. For example, an induction period is necessary for a pre-catalyst to undergo a transformation to form the active catalyst, before it can take effect in a catalytic reaction. In an auto-catalytic reaction, the rate of reaction remains slow until sufficient products have formed to catalyse the reaction. Heat is normally used to accelerate reactions.

In the case of an exothermic reaction the rate of the reaction is initially very low, heat is generated by the reaction and the rate of reaction increases. A time interval is needed for the reaction to reach full reactivity. This time interval is regarded as the induction time. Insufficient induction times in coatings can cause coating failures. For example, many epoxy coatings require a fixed time interval in order to make the epoxy resin and hardener fully compatible.

When this pre-reaction time is not met, it will produce a poor film appearance and poor coating performance. Induction time will usually increase with low temperatures and/or high humidity.

In coating application improperly inducted epoxies will lead to:

- A reduced gloss
- Polyamide/amine blushing
- Poor colour appearance
- Lower corrosion and chemical resistance
- Poorer film.

Wet Film Thickness

Wet film combs are available as plastic disposable types and a reusable metal comb. A plastic wet film thickness comb is a "use once" type gauge as the plastic wet film comb can distort with excessive use. To use the comb is not difficult. Make sure the "teeth" are clean and free from any previous paint coatings then immediately as the paint has been applied place the comb perpendicularly into the wet film.

The standard for this test can found in the ASTM D 4414-5 "Standard Practice for Measurement of Wet Film Thickness by Notch Gauge".

Wet Film Thickness Comb

The two end teeth will penetrate down to contact the underlying surface and will be wetted with coating. Withdraw the gauge from the wet film and read the highest wetted step. If all of the teeth contain coating rotate the gauge to a higher WFT range. The gauge must be wiped clean of coatings after each time it is applied into the wet film. When applying the protective coating by spray application, gun technique is important to provide an even, continuous wet film.

If the gun is held too close to the surface then over application will occur like sags, runs, even pinholes with air being blown into the coating creating these voids. If the spray gun is held at too far a distance from the object being painted, dry spray could occur. This happens when the paint particles leave the nozzle orifice and before the paint reaches the surface, the solvent will evaporate from the paint, resulting in a "dry sand paper like" texture on the surface.

Each spray pass should overlap the previous pass by 50% so even coverage is achieved; this is essential as not to create any holidays and missed areas. The correct method of spraying is triggering the gun at the start and the end of each pass or stroke. Smooth coverage and a continuous paint film will be achieved with the correct spraying technique.

The Formula for DFT/WFT/Volume Solid Calculations:

- Wet Film Thickness: (WFT)
- Dry Film Thickness: (DFT)
- Volume Solids (%): (VS)
- WFT = 100 x DFT ÷ VS which can be rearranged to: DFT = VS x WFT ÷ 100.
- **Example 1**: In order to achieve a DFT of 90 microns using a paint with 60% volume solids, we would calculate: WFT = 100 x DFT ÷ VS WFT = 100 x 90 ÷ 60 WFT = 150 microns: 100 x 90 = 9000 ÷ 60 = 150
- **Example 2:** Say we add 0.5 litres of thinner to every 5 litres of paint with a volume solids of 60% calculate the new Volume Solids: newVolS% = VolP x oldVolS% (VolP + VolT) newVolS% = 5 x 60 (5 + 0.5) newVolS% = 54.5%
- **The Formula**: The volume solids will change with the addition of solvents or thinners. We must then re-evaluate the volume solids before we calculate the wet film thickness. The added solvents change the original volume solids and a recalculation is therefore necessary. This assumes of course you know the exact amount of solvent added. The added solvent changes the original volume solids and therefore you must know the exact amount of solvent added.

- The calculation for volume solids is: volume solids/100 + % of solvent = adjusted volume solids.

Electrostatic Spraying

Electrostatic spray painting involves the concepts of charge and electric fields. Fundamentally the electrostatic technique of applying a coating to a metal surface involves the law of attraction between positively and negatively charged particles. Compressed air forces the paint through the end of the spray gun, which atomises the liquid into a fine spray. Atomisation essentially breaks up the paint into small droplets and this process is also utilised in electrostatic spraying, but with one unique difference regarding the methodology.

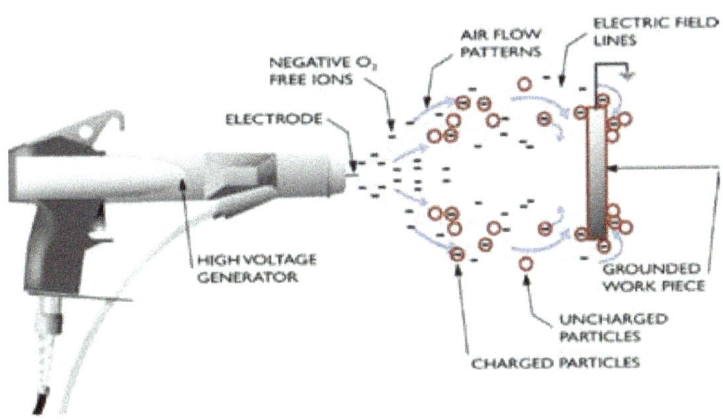

Electrostatic Spraying

The paint is atomised in a static field which is formed at the end of the electrostatic spray gun. Just before the mist of paint leaves the nozzle, it is given a positive charge. The charged paint droplets are sprayed through a strong electric field which is a term used to describe patterns of forces. The negatively charged, grounded metal item attracts the positively charged liquid to its surface very much like a magnet.

Here, a basic law of electricity is inherent in the electrostatic system of spray applying paint, namely that opposite polarities attract. Scientifically, this relatively simple concept is known as **Coulomb's Law** which states that the same electrical charges repel each other whereas opposite charges attract as in iron filings to a magnet.

A Corona discharge is crucial for attraction within the electrostatic spraying technique to take place. This occurs through the ionisation of the paint within the electric field surrounding the conductor (in this case the metal). This is known as '**Corona Discharge**'. Even though the corona is gaseous, it is still conductive because it contains free electrons and therefore enables the current to flow from the charged gun through the air.

The electric field also influences the force of attraction. After the electric field has been created, the ions follow the field lines towards the negatively charged surface. The force attracting each of these positively charged particles to the grounded metal substrate is equal to the charge of the particle itself multiplied by the electric field strength. Even though the electrical charge given to the paint is small, it is strong enough to counteract the effects of aerodynamics and gravity.

In fact, the force of attraction is up to 75 times greater than the force of gravity alone and when the particle has reached the metal, the force of attraction is strong enough to keep it there. This is partly because the chemicals generally used in electrostatic painting are strong dielectrics in that once the particles become charged, the charge remains for a sufficient amount of time for the attraction to be sustained.

As the attraction of the electrically charged particles to the grounded surface is so strong, paint drift is reduced. This means there is a high transference rate of paint to surface averaging at 98%. Furthermore, the paint droplets are pulled towards surfaces in all directions, so that all sides of target areas are coated. Some particles actually change direction as they are pulled up, down and sideways towards all angles of a facade.

Advantages of the Wrap around Effect

- The material is electrostatically charged as it passes through an electrostatic field produced between the electrode on the front of the gun and a grounded object.
- The charged particles of the material are attracted to the grounded (neutral) object and form an even coating.
- The charged material then wraps itself around the object, increasing the surface space that is coated.

The under-side of the substrates are coated and paint is renowned for reaching around the back of surfaces if an electrostatic system is used. The new coating will even get into crevasses and this **'Electrostatic Wraparound'** makes it an ideal technique for coating hard to reach areas. The wraparound effect that electrostatic painting is well-known for also allows this technique to be used in enclosed spaces such as offices and retail outlets.

Less paint is released into the air because it is so strongly attracted to grounded surfaces, so fumes are reduced and the environment is safer than when other painting methods are used. The particles leaving the spray gun have the same positive charge and therefore repel each other.

This means that they do not conglomerate into large droplets, so the paint mist remains fine and evenly sized. Likewise, the attraction of spray to uncoated surfaces is stronger than those that have already been coated. It is as if the charged paint droplets seek out the uncoated areas of substrate resulting in a more uniform coverage than other more traditional methods of painting.

Solvent-borne coatings, including low, medium, and high solid coatings, enamels, lacquers, and two-component coatings, are often applied with air spray or air-assisted airless technology.

- **Air Spray Electrostatic Spraying** uses a low-pressure fluid stream that is mixed with compressed air at the air-cap to atomize material in a controlled manner. It is used for the application of low to medium viscosity fluids for products requiring a high quality, Class A or decorative finish.
- **Air-Assisted Airless Electrostatic Spraying** uses a high-pressure fluid supply for atomization and compressed air at the cap for pattern control. Air-assisted airless electrostatic spraying solves many problems that arise with the use of high-viscosity and high-solids coatings, and other issues associated with heating and using higher fluid pressures to aid in the atomization of more viscous materials.

CHAPTER NINE
Metallic Coatings

Hot Dip Galvanizing

When a clean iron or steel component is dipped into the molten zinc (which is commonly at around 450°C) a series of zinc-iron alloy layers are formed by a metallurgical reaction between the iron and zinc. The rate of reaction between the steel and the zinc is normally parabolic with time and so the initial rate of reaction is very rapid and considerable agitation can be seen in the zinc bath. The main thickness of coating is formed during this period.

Subsequently the reaction slows down and the coating thickness is not increased significantly even if the article is in the bath for a longer period of time. The zinc-iron alloy reaction layers consist of a layer of zinc with iron reaction compounds and an outer covering of zinc of the same composition of the bath. The usual galvanized coating consists of three distinct iron-zinc alloy compounds (namely **Zeta, Delta and Gamma**), formed by the metallurgical reaction between iron and zinc. These three layers are covered by an outer layer of free zinc named **Eta**.

- **Eta Layer** – 100% zinc outer layer
- **Zeta Layer** – 94% zinc and 6% iron
- **Delta Layer** – 90% zinc and 10% iron
- **Gamma Layer** – 75% zinc and 25% iron.

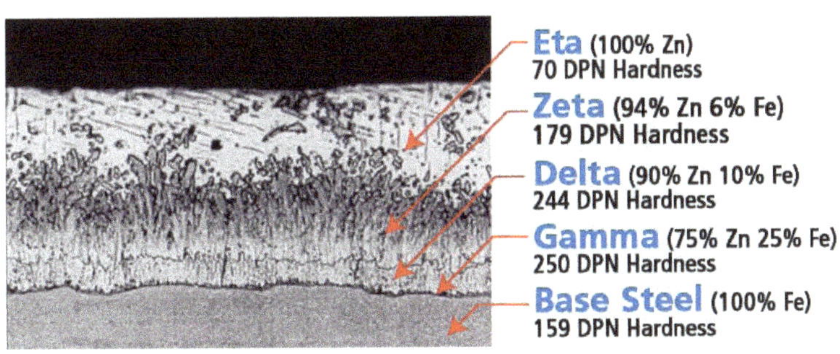

The whole galvanizing process includes surface cleaning, rinsing and pickling the components prior to the hot dip to remove surface contaminants such as dirt, oil, grease and also to dissolve surface oxides that may have formed.

The Galvanizing Process Steps are:

- Surface Preparation
- Fluxing
- Dipping
- Post Dipping Treatments
- Inspection.

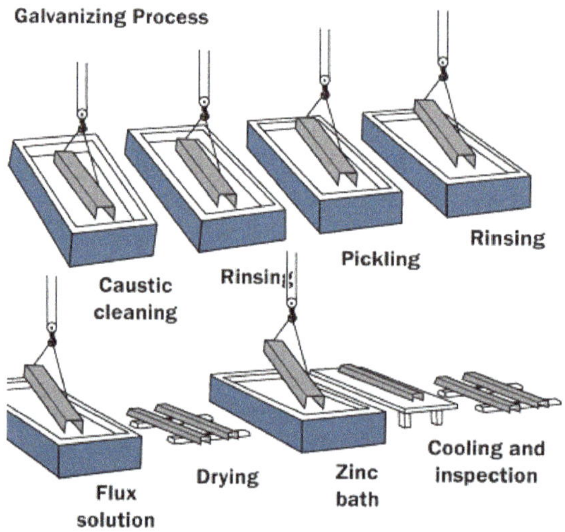

Zinc Galvanizing Process

Surface Preparation

The structure to be hot dipped galvanized must be free of welding slag, mill scale, paint, oil and grease as well as other contaminants. Cleaning is usually a two-step process operation. The work piece is first dipped into a hot caustic bath or similar solution to remove oil, grease and other contaminants. Next the work piece is thoroughly rinsed with clean water and immersed in a hot mineral acid bath, usually sulphuric or hydrochloric acid, to remove rust, mill scale, and other inorganic contaminants.

Sulphuric acid baths are always heated to 60°C – 79°C. Hydrochloric acid solutions are normally used at ambient temperatures of 24°C – 38°C to prevent excessive fuming. Occasionally the work piece has to be cleaned by dry abrasive blasting due to the nature of the work piece. Sand casting for example must be abrasive blast cleaned to remove the burnt-on sand from the casting operation prior to galvanizing.

Fluxing

During the cleaning and handling process, oxides may form on the work piece. These oxides must be removed by dipping the work piece in either an aqueous pre-flux solution (Zinc Ammonium Chloride) or by passing the work piece through a molten layer of dry flux on the surface of the molten zinc in the kettle.

Dipping

The work piece is then dipped at a very controlled rate into a bath of molten zinc. The rate of immersion and withdrawal is critical, depending on the type of metal alloy and the thickness of the metal being dipped. Objects larger than the bath can be galvanized by a procedure referred to as double dipping, first dipping one end then the other. Small articles, including nuts and bolts, can be placed in perforated baskets and then dipped.

Strip steel sheets can be continuously galvanized very quickly. Sheet steel can be dipped singly in a rapid sequence. The work piece remains in the molten zinc until the alloying of the zinc/metal surface is complete. This can be from as little as 30 seconds to 5 minutes, but it can be longer for heavy articles. Generally, for the alloying process to be complete, the work piece must attain the temperature of the zinc bath, which may vary from 450 °C to 475°C.

Post Dipping Treatments

When the alloying is complete, the work piece is removed from the zinc bath and cooled either by blowing with compressed air or dipping the work piece into a water quench tank. If the work piece is excessively rough, it may be rolled, wiped, centrifuged or air blasted to smooth the molten zinc prior to cooling or quenching. The properties or appearance of the galvanized metal can be improved by such treatments as chroming, phosphating, light rolling, and roller levelling.

Inspection

At each step during and after the hot-dip process, the work piece should be inspected to ensure compliance with the specification. Visual inspection and a variety of simple laboratory and field tests can be done to determine thickness, coating uniformity, coating adhesion and appearance. In addition, embrittlement and defects resulting from fabrication and design are considered during inspection.

The thickness of the zinc deposited during the HDG process is affected by several items, including the speed of the dipping and the speed of the withdrawal from the zinc bath, and the steel chemistry itself. The reactivity of the silicon and phosphorous content of the steel can cause excessive thickness and roughness of the galvanized zinc layer.

Thickness of galvanizing is generally specified in ounces per square foot, rather than in dry film thickness. The coating thickness can be measured in one of four ways, magnetic thickness gauges, stripping, weighing before and after galvanizing, and optical microscopy.

The most commonly used method is the magnetic thickness gauge, as shown in ASTM–E–376. These may be manual magnetic pull off gauges (SSPC–PA2 Type 1) or its electromagnetic fixed probe gauges (SSPC–PA2 Type 2). The accuracy of these gauges is normally within 3% to 10%. Accuracy can be affected by surface roughness of the galvanizing. The gauges should be calibrated at least once per working shift to maintain accuracy and repeatability of the measurements.

The stripping method is detailed in ASTM–A–90. It is a destructive test in which a sample is first weighed after galvanizing, and then the galvanizing is stripped from the steel surface by dipping in a chemical solution and the sample is weighed again. The difference in the two weights is the weight of the coating. This method tends to give a slightly higher coating thickness than the actual value. Weighing an article before and after dipping gives a weighted value that can be converted to dry film thickness by mathematical calculations. This method may underestimate the coating thickness by as much as 10%.

Measurement by optical microscopy is done in accordance with ASTM–B–487. The coating thickness is measured by cutting out a portion of the coated steel, mounting it, and preparing the edge of the mounted piece for optical microscopy. The ASTM test method details the conditions that must be met to obtain an accurate reading of the coating thickness. This is a destructive test method normally used in laboratory conditions and may not be appropriate for all galvanized articles.

Design Considerations

It is important that the design of the structure is considered before considering a galvanized process.

- Skip weld and crevices may trap acid which causes damage to the galvanizing as it could vaporise during the process or stain the coating at a later date.
- Ensure there are no areas where air pockets or air bubbles may form preventing full contact with the molten zinc.
- Ensure the structure is of a size that can fit into the zinc tank.
- Warpage of the article is a possibility due to unequal thickness of the article or structure.
- The article or structure should be made of the same alloy.
- Faying surfaces should not be galvanized as this would have to be removed later.

Finished Appearance

Some specifications call for a specific appearance of the galvanized article such as brightness and roughness which are controlled by the steel chemistry, temperature of the zinc bath, speed of dipping and speed of withdrawal, as well as additives in the zinc bath such as aluminium. After several years of atmospheric service, the galvanizing will become a relatively uniform grey colour, therefore the original grey colour is not normally a cause for rejection.

Hot Dipped Galvanized Surface

Surface roughness is not normally a cause for rejection unless the roughness interferes with the intended use of the dipped article. The rate of withdrawal can cause zinc spikes below the bottom of the article. Although these are not normally a cause for rejection, they can be a safety hazard handling the dipped article. Bare areas of hot-dipped galvanizing are not normally acceptable, unless intentionally left bare as part of the fabrication process. Unintentional bare areas can be repaired with zinc rich coatings and spray metallizing.

Causes of Bare Areas:

- **Mill Scale/Rust Scale** – If this scale developed during the manufacturing process and is not removed, the molten zinc may not adhere to these locations.
- **Residues** – If the structure still has residues on the surface such as paint, grease, oil etc. the molten metal will not adhere and you will get a localized grey/black spot.
- **Rolling Defects** – If the structure has steel rolling defects such as laminations, shelling, bare spots on the galvanized surface will occur.
- **Over-drying** – If a pre-fluxed structure is left too long before galvanizing you may encounter bare spots.

Bare Spot/Rusting

Adhesion

Adhesion of hot-dipped galvanizing is normally very good due to the alloying process that takes place during the hot-dipping. Adhesion can be tested by using the point of a knife (paring test) and attempting to pry the coating from the steel surface.

The removal of small particles by paring does not indicate adhesion problems. ASTM–A–123 specifies a pivoted hammer test in which a riveting hammer modified to weigh (210 g) is used. The hammer is allowed to fall from a horizontal position from its own weight at a 90° angle onto the galvanized surface. Ability to remove or lift the coating between the impressions constitutes unsatisfactory adhesion at the locations.

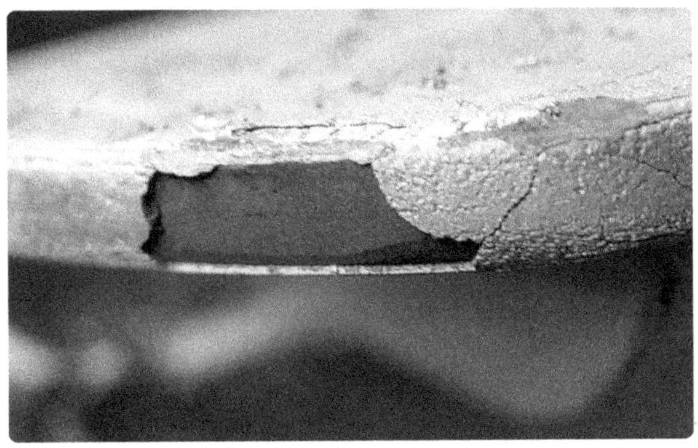

Loss of Adhesion/Cracking

Pre-treatment prior to coating

There are four recognized methods of surface pre-treatment that produce a sound substrate for the duplex coating. These methods include T-Wash, Weathering, Etch Primer and Sweep Blasting. When paint is applied over a hot dipped galvanized surface, the resulting combination is known as a duplex coating.

The service life of a duplex coating generally exceeds the sum of the separate lives of the component parts. The first decision is should galvanizing be over-painted and whether painting is necessary. The decision is best made during the design stage.

There are two main reasons for painting galvanized surfaces:

- Increase the life of the coating.
- Offer colour and aesthetics.

The important aspect to over painting the galvanizing is achieving good adhesion. Without the good adhesion, paint will fail prematurely, leading to flaking and peeling.

Four methods of surface pre-treatment:

- Mordant solution (T-wash).
- Allowing the surface to weather for a minimum of 12 months.
- Etch primer.
- Sweep blasting.

T-Wash – This process has been used successfully for a number of years. Newly galvanized steel is thoroughly degreased and then treated with the T-wash. The T-wash reacts with the galvanizing to form a surface onto which a paint coating can be applied. The reaction is a visual one in that a successful application of T-wash produces a black surface colouration. Avoid ponding. Apply as per supplier's application data sheets.

Weathering – The weathering process becomes fully effective after a galvanized surface has been exposed to the atmosphere for a period of approximately 12 months. Preparation generally consists of removing loose adherent materials by using abrasive pads or stiff brushes. A detergent and freshwater wash is then conducted. The surface must be fully dried before application of the coating.

Etch Primer – The major disadvantage of etch primers, is the absence of any visible colour change after the application as in the case of T-wash. Therefore, there can never be complete confidence that all surfaces have fully reacted to the primer. Both single and two-part etch primer are available; however, best results are found with the two-part. Always check the application data sheets on pre-cleaning, application and environmental criteria.

Sweep Blasting – Sweep blasting can be conducted at low pressures with the use of fine grade non-metallic abrasives. It is a process which can be very effective, but only when undertaken with a high a degree of expertise. Excessive blast cleaning must be avoided as this can remove the zinc coating.

Liquid Metal Assisted Cracking

Liquid Metal Assisted Cracking is a form of cracking that may occur when steel components come into contact with molten zinc. This form of cracking is uncommon but if it is not detected and repaired it can have serious consequences on the performance of the structure. LMAC is a phenomenon that may occur when steel components are hot dipped galvanized, but it is important to note that this form of cracking can only happen when the steel is in contact with molten zinc.

Liquid Metal Assisted Cracking

LMAC may be characterized by a crack through the entire cross section that is clearly evident when the component is lifted from the bath. It is also possible that the crack will not have travelled through the whole cross-section, will have filled with zinc, and the zinc will have uniformly deposited over the surface of the steel, rendering the crack invisible to visual inspection. Nevertheless 100% inspection is recommended.

It is generally accepted that there are three prerequisites for LMAC to occur:

- Stress Levels
- Material Susceptibility (where the local stress is greater than the yield stress)
- Liquid Metal.

Repair of LMAC

Most cracks in a galvanized component can be repaired using an approved welding repair procedure. During the repair the procedure should ensure that the crack is completely removed and Magnetic Partial Inspection (MPI) should be used to ensure that this is the case. The grinding should continue approximately 50 mm (2 inches) back from the crack. The repair should then be touched up, either with a zinc rich paint or zinc metal spray. The latter will require a high level of surface preparation.

Advantages of Hot-Dip Galvanizing:

- Excellent long-term service without topcoats in environments between pH 5 and 10
- Excellent protection of complex, fabricated structures
- Structures can be coated internally and externally at the same time
- Structures ready to ship and install immediately after hot-dipping.

Disadvantages of Hot-Dip Galvanizing:

- Length and width of hot-dipped structures is limited by the size of the bath
- Top coating requires sealing of porosity first, or bubbling and blisters can occur
- Galvanizing must be top coated for service in environments below pH 5 and above pH 10 otherwise they will sacrifice too rapidly
- Galvanizing is not effective in immersion conditions, zinc sacrifices itself too quickly.

CHAPTER TEN
Thermal Sprayed Coatings

Thermal spray is a generic term applied to several processes that deposit finely divided metallic or non-metallic materials in a molten or semi-molten state onto a properly prepared surface. Thermal sprayed coatings function as barrier coatings, which isolate the substrate from the environment, and as galvanic coatings, which corrode in preference to the underlying substrate.

A build-up of the metallic coating corrosion products gives added protection to the substrate. When used in conjunction with typical cathodic protection systems, thermal spray coatings reduce current demand. For maximum protection in atmospheric environments, thermal sprayed coatings may be sealed with an organic top coat.

A relatively thick metal coating without a sealer or top coat provides the longest maintenance interval. While the addition of a sealer coat or top coat may extend the service life of the metallizing, sealed or top coated systems will require maintenance primarily for cosmetic purposes.

Thermal spraying can be accomplished with different types of equipment:

- **Flame Spray** – (usually acetylene or propane gas)
- **Electric Arc Spray**
- **High Velocity Oxygen Fuel (HVOF).**

Flame sprayed metals are normally supplied in wire form with a ⅛ inch or $^3/_{16}$ inch diameter. Ceramics and plastics are commonly supplied in powder form, which requires either a hopper container feeding the powder by gravity into the flame or a separate low-pressure feed line into the flame. Electric arc spraying equipment gives the higher productivity, up to four times more than flame spray. Plasma sprayed metals and ceramics are normally supplied in powder form.

Surface Preparation Requirements

Thermal sprayed zinc (TSZ), aluminium composites, and thermal sprayed aluminium

(TSA) require a minimum of ISO Sa 2.5/Nace 2/SSPC – SP10 Near white metal blast cleanliness for atmospheric service. Best results are obtained over ISO Sa 3/Nace 1/ SSPC – SP5 White metal blast cleaning. Surface profile requirements are normally in the 50μm to 125μm range, but some owners require 150μm. The surface profile must be angular. The blasting media may be a metallic abrasive or an aluminium oxide abrasive.

Characteristics of Gas Flame Sprayed Coatings

The gas flame process has been around since the 1900s and it is the oldest metallization technique to date. Flame sprayed metallic particles (aluminium or zinc) achieve a roughly spherical shape as the molten metal leaves the head of the spray gun. Upon striking the surface, the particles flatten into irregular shapes (splat) and patterns that flow into the pores and irregularities of the substrate; the result is a relatively rough, porous surface.

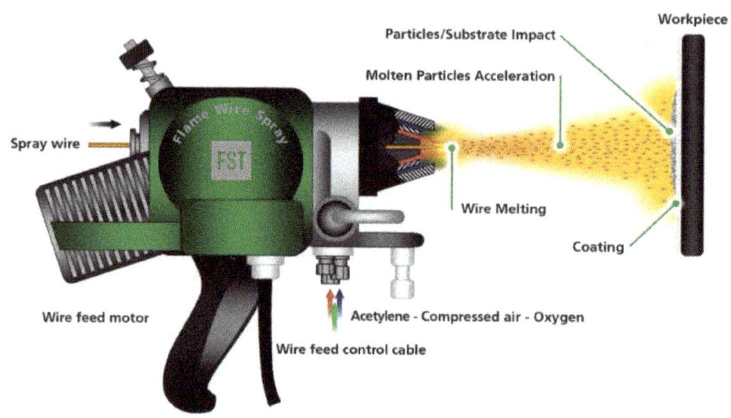

Gas Flame Spray

In the gas flame procedure a wire is fed by a driven roller system through the centre of an oxygen propane flame where it is melted. An annular air nozzle then applies a jet of high-pressure air, which atomizes and projects the molten material onto the work piece. The driving of the wire is typically via an air motor and gearbox that forms part of the pistol. Wire diameters that can be flame sprayed as standard range from 1.6mm to 4.76mm (1⁄16" to 3⁄16").

Wire is typically dispensed from coils or production packs or drums. The wire is fed through the wire drive mechanism where it is mixed with the compressed air and fuel, usually acetylene and propane, and projected onto the substrates. The adhesion is achieved by both mechanical and chemical means into the anchor pattern provided by the abrasive blast cleaning of the substrate.

Pre-heating the substrate immediately prior to application of the flame sprayed coating improves adhesion. It is said to remove any remnants of surface moisture and expand the metal slightly, both of which should lead to better wetting of the substrate by the molten coating. Common preheat temperatures range from 35°C to 65°C. Pre-heating is not normally done with the arc spray equipment.

Thermally Sprayed Aluminium – Thermal sprayed aluminium (often referred to as TSA) has obtained an extensive track record in corrosive environments such as offshore platforms, piping and structures. As aluminium has a melting point of 660°C (1220°F) it is used on hot structures such as flare booms and piping.

TSA Application by Gas Flame Spray

Thermal spray aluminium has found a popular place on new build projects which are subject to hostile environments and locations where maintenance can be difficult. Although the initial application cost of thermal spray aluminium is higher than paint coatings, the annualised cost will be lower over the life of the structure (20 + years).

Thermally Sprayed Zinc – Zinc coatings are either pure zinc or an alloy of zinc and 15% aluminium. Sprayed coatings of pure zinc have been available since the 1920s and are typically applied with a thickness of 100 microns or more. If not further coated, the pure zinc coating will form a stable, passive film of zinc corrosion products (zinc oxide, zinc hydroxide, zinc carbonate) with a predictable weathering rate. In many environments the sprayed zinc alone can provide sufficient service life.

The zinc 15%-aluminium thermal spray coating was introduced in the late 1970s and is about the highest aluminium composition in zinc that can be industrially made into wire. Because the micro-structure of each of the droplets of this sprayed coating consists of both a zinc-rich and aluminium-rich phase, Zn–15%Al offers the optimum corrosion protection to steel in aggressive atmospheric conditions such as marine environments.

Characteristics of Electric Arc Spraying

Electric arc spraying allows a higher deposition rate, with improved economy and uses two consumable zinc wire electrodes. The wires are given opposite electric charges and are fed close together to maintain an electric arc, causing the wires to melt. The molten material is sprayed onto the prepared surface with compressed air.

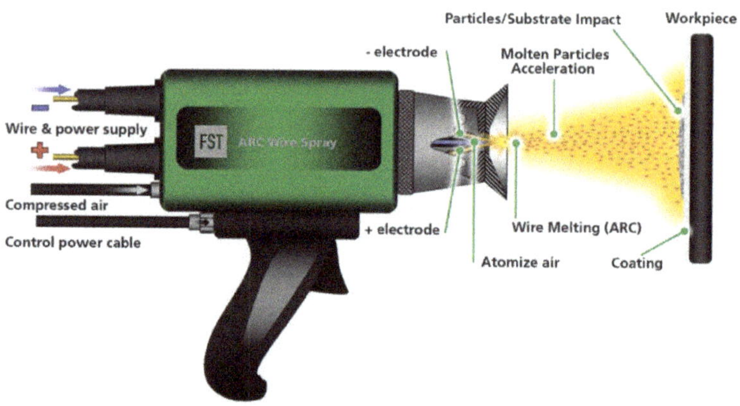

Electric Arc Spraying

This is important when considering atmospheric conditions that will dictate the protective thickness required, as a very long life or highly corrosive conditions provide a thicker coating. Edge coating, a normally difficult task for all coating systems, is easily and efficiently achieved using the zinc thermal spraying process. The use of sealers and/or topcoats over a zinc metalized coating will further protect the surface and usually provide an even longer life of the coating system. Again, environmental conditions will determine the use and type of sealers.

Characteristics of High Velocity Oxygen Fuel (HVOF)

Thermal spray processes include High Velocity Oxygen/Fuel (HVOF) systems, which produce very dense coatings. Powdered alloy feedstock is fed into a chamber and accelerated toward the substrate at very high velocities. The process involves injecting a fuel gas with oxygen into a combustion chamber. The combustion gases are discharged through ports in the combustion head, increasing the gas velocity to hypersonic speeds.

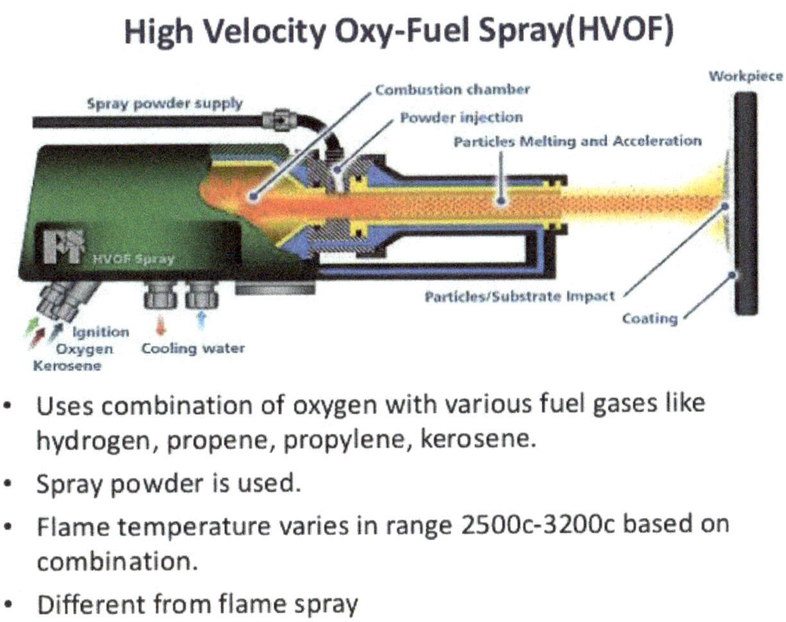

Powder is injected into the intersection of the discharge ports by using a carrier gas such as argon. Flame temperatures are relatively low (2900°C). The high velocity and low application temperature require a finer-particle powder and tighter particle-size control than other thermal spray techniques. While conceptually similar to combustion spray technology, it must be emphasized that the resultant coating is substantially different.

The ability to apply the molten product at hypersonic speeds produces very dense coating on the substrate without any of the porosity common to conventional thermal spray. In many applications HVOF can be substituted for plating applications such as chrome used for hard wearing surfaces.

Advantages of Thermal Sprayed Zinc and Aluminium

Long service life – The tension legs on many platforms in the North Sea were originally designed to provide a minimum of 20 years' maintenance free service. Repeated in-house coating surveys now indicate that the service life will be at least 40 years due to having a thermally sprayed coating applied.

Rapid Turn-around – Structures can be shipped within hours of final application and quality control approval. Sealer coats can be applied immediately after the TSA or TSZ has been applied. Top coats can be applied within 4 hours. When sealer and top coats are not required, the TSA or TSZ coated structure can be shipped immediately.

Resistance to Mechanical Damage – TSA and TSZ have excellent resistance to mechanical damage showing relatively little delamination in even the worst conditions. Damaged areas can be spot repaired with either itself or organic primers such as zinc rich epoxy.

Low VOC emissions – TSA and TSZ emit zero VOC's. The applicator's problems with collection, storage, and disposal of hazardous waste are significantly reduced.

Disadvantages of Thermal Sprayed Metals

Over-spray – Overspray during application tends to be in the form of dust, which is potentially explosive in the presence of sparks. TSA dust, in particular, can burn with an aggressive flame. From a health and safety aspect, several problems or hazards are caused through the dusts produced while metal spraying. The quantities of dust involved are very much dependent upon the efficiency of deposition of the process.

It is estimated that approximately 30% of metal wire is turned into respirable dust. Dusts can pose a hazard because of a combination of particle size and concentration in the immediate atmosphere. With a high concentration of dust it is possible to create an explosive environment in some circumstances. Extraction is the method most usually used to maintain dust levels below the explosive limits.

The extraction must be sufficiently efficient to ensure all dust particles are removed from the air during operations. The quantities required to be extracted should be calculated using the efficiencies of the spraying as a guide. Dust built up in the extraction ducts is a particular hazard which should be considered whilst designing the equipment layout.

Handling of dry dusts causes some problems when extracted dusts are collected in a hopper or chamber. The collection areas will have a high concentration of dusts. The filters should be specifically developed to be used with metal coatings. Wet dust collection systems can be used especially for fixed location. Dust debris must be carefully and safely disposed of in accordance with local waste regulations.

Zinc flu – A temporary illness known as zinc flu, similar to symptoms of a heavy cold, can affect operators who breathe in large amounts of zinc dust or fumes. Wearing the appropriate respirators can eliminate this problem.

Operator expertise – Operator expertise must be very consistent. The operators can apply only 50µm – 70 um per pass and usually make 3 to 4 passes to achieve the specified DFT. This requires careful overlapping of spray passes to avoid uneven films or holidays.

Adhesion – As the coating bonds by mechanical strength, the adhesion value of the sprayed metal is important. The specification will detail the minimum adhesion values; however, in general the minimum adhesion values should be:

- Gas application – 600psi/4.05mpa
- Arc application – 1000psi/6.76mpa

Adhesion tests are generally conducted on test plates at the start of the shift and then on the structure if adhesion values do not meet the minimum requirement of the test plate. There are various adhesion standards used globally but most will use a dolly which is glued to the surface and pulled off (pull-off gauge) to demonstrate the adhesion values. A skilled coating inspector/supervisor should perform this test using the appropriate standards.

Sealers and Topcoats

Aluminium and zinc thermal sprayed coatings and their alloys have porosity in the applied film up to 15%. This porosity can extend all the way down to the substrate. Sealing of the thermal sprayed zinc or aluminium extends the service life significantly, particularly in high humidity, splash zones and immersion conditions.

This is usually accomplished immediately after the thermal spray is completed in order to avoid any adhesion problems to the oxides that develop from reaction of the zinc or aluminium with the atmosphere. These sealers were originally thin coats of either aluminium pigmented silicone or thin coats of vinyl.

Problems with visibility with the remotely operated vehicles (ROV) used to survey the underwater sections of offshore platforms has led to an increasing use of a thin coat of yellow epoxy as a sealer coat. For service temperatures higher than 120°C aluminium pigmented silicone sealers are preferred.

Other sealers that have been used include air drying phenolics, silicates. It is important that the sealer coat only seals the thermal sprayed coating, leaving a very thin film on the surface. Thick coating films would reduce the cathodic protection capability of the thermal sprayed coating.

Advantages of sealing are:

- The sealer penetrates the pores
- The sealer smoothes the surface texture
- The sealer can be coloured to add an aesthetic finish and is easier to keep clean
- Topcoats such as epoxy or urethane acrylic can also be used if a high-quality finish is required.

CHAPTER ELEVEN
Inspection Instruments and Practices

The Bresle Test - ISO 8502

This standard is intended for use in connection with the determination of several types of contaminants on the steel surface. So far the standard has been used for determining conductivity and chloride concentration on blast cleaned steel before painting application. When a sample is to be taken according to ISO 8502 a small adhesive plastic patch with a latex membrane is used where the area and volume are known.

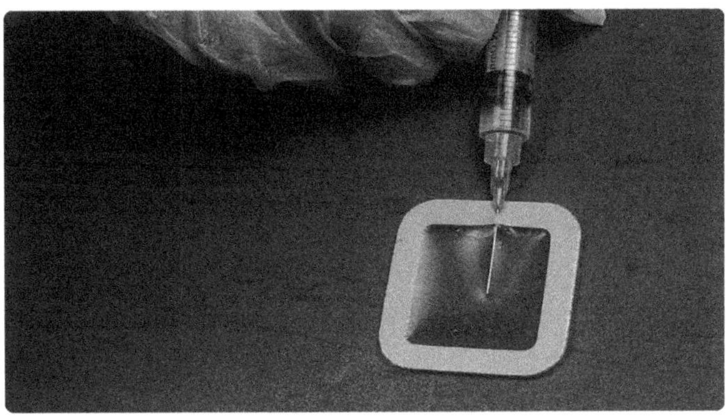

Bresle Test

First, we remove the protective tape from the adhesive back of the patch and the cut out in the patch, and then we attach it to the blast cleaned steel. When doing this make sure that as little air as possible is trapped in the analysis compartment in the patch. 15 ml of distilled or deionized water is poured into a graduated measuring tube. The conductivity of the distilled water is measured using a calibrated conductivity meter (note the value).

From the graduated measuring tube 3 ml of distilled water is drawn using the needle of the syringe. The 3 ml of water is injected into the analysis compartment through the edge of the Bresle patch. The water is pumped back and forth between the patch and the syringe 2–4 times per minute for about 5 minutes. All the water is finally drawn into the syringe and transferred back into the graduated tube with the 12 ml of water.

The conductivity probe is inserted and the conductivity measured. In order to calculate the salt concentration in mg/m^2 the actual increase in conductivity in uS/cm is multiplied by a factor of 6.

Dry Film Thickness Instruments

Magnetic DFT gauges can be either type 1 manual pull off gauges or type 2 electromagnetic positive pressure probe gauges according to "SSPC-PA 2 Method for Measurement on Dry Film Thickness on Ferrous Substrates".

Other standards that may be used include ISO 19840 – Paints and Varnishes – Corrosion Protection of Steel Structures by Protective Paint Systems – Measurement of and Acceptance criteria for the Thickness of Dry Films on Rough Surfaces and the ASTM D 1186.

Electromagnetic DFT Gauge Type 2

Type 1 gauges are available as dial type 1A Banana Gauges. Accuracy of measurement ranges from 1% to 5 % on type 1A gauges. Type 1A – DFT gauges may be calibrated or adjusted on small chrome plated steel panels such as National Institute of Standards and Technology (NIST) plates available from the various companies. These are available with certified plates ranging from 0.5 to 80 mils. Proprietary calibration gauges are available from companies like Elcometer in the United Kingdom.

Calibration/Adjusting of the gauge is quite simple. Place the gauge on the plate that is closest to the expected thickness of the coating being measured and engage the magnetic tip. Rotate the dial on type 1A until the magnetic tip is disengaged. At that point read the DFT.

Actual use of the type 1 gauges requires measurement of the bare steel to obtain what is called "Base Metal Reading" prior to actually measuring the painted areas. The gauge is placed on the bare steel surface and the measurement is taken exactly the same as on painted steel, but the reading is quite low, usually less than 25μm. This value is recorded and used for final DFT measurement.

Dial Type 1A Banana Gauge

At this point, the gauge is used on the painted steel and the DFT measurement that is obtained has to be adjusted for accuracy. The first step is always the deduction of the BMR from the actual reading. Next an adjustment must be made for the calibration correction. If the gauge reads higher than the calibration plate on which it was tested, the difference between the plate and the calibration reading must be deducted from the actual reading.

If the gauge reads lower than the calibration plate on which it was tested, the difference between the plate and the calibration reading must be added to the actual reading. The calibration of the type 2 electromagnetic gauge is easier and simpler. The manufacturer normally supplies a small pack of plastic shims (like the ones pictured) of a known thickness with the gauge.

The shim that is closest to the expected DFT of the coating measured is selected and placed on the bare steel. The gauge is placed on the plastic shim and a reading occurs automatically. The gauge can then be adjusted by pressing on the plus or minus buttons until the gauge reads the same as the plastic shim.

Zero Plate with Various Shims

Eddy Current type gauges look exactly like type 2 electromagnetic gauges but are not covered by SSPC–PA 2. They are covered by ASTM D 1400 "Standard Test Method for Non-destructive Measurement of Dry Film Thickness of Non-conductive Coatings applied to a Non-ferrous Metal Base". These gauges measure DFT on non-ferrous substrates, such as stainless steel, aluminium, copper and brass. Their accuracy is also 1% to 3%. The calibration process is similar to type 2 gauges but varies according to the manufacturer's instructions.

Some manufacturers suggest calibration at both zero on the bare non-ferrous metal and at the expected DFT of the coating being measured until both measurement points are accurate. Other manufacturers require only single point calibration similar to type 2 gauges.

The important point is that regardless of whether single point or dual point calibration is performed, it must be performed on the non-ferrous metal coated surface. A calibration on aluminium is normally performed at the manufacturer's facility. If the gauge is being used to measure non-conductive coatings on a stainless steel or other non-ferrous metal, then it should be calibrated on that metal.

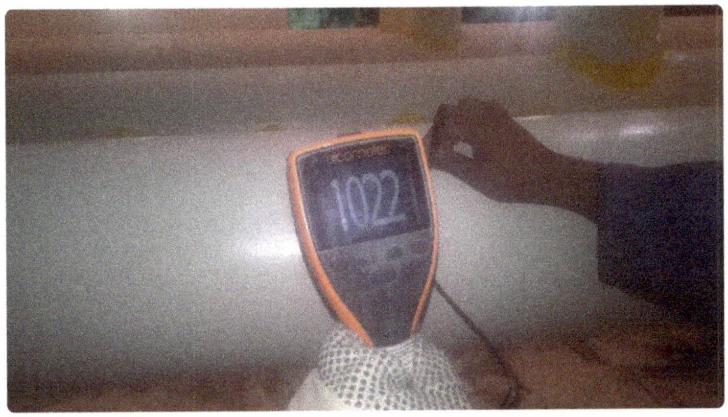

DFT Measurements with a Type 2 Electromagnetic Gauge

Adjustment of the Instrument

- The gauge should be in good working order and correctly adjusted. It should be verified on uncoated test panels at zero with shims/foils above and below the specified dry film thickness.
- **Adjustment:** Adjustment is carried out by the user, usually the painting contractor or inspector.
- **Calibration:** Calibration is carried out by the gauge manufacturer using traceable thickness standards in most cases.

Measurement

- Measurement on the dry film thickness shall only be taken after adjustment and checking of the instrument has been carried out.
- The instrument shall then be used in accordance with the instrument manufacturer's instructions.
- Following completion of a series of measurements and preferably during the measurements the adjustment of the instrument shall be re-verified.
- When during a series of measurements an individual dry film thickness value does not meet the criteria a repeated measurement not more than 10mm from the point of the first measurement shall be carried out.
- The first value shall then be rejected and replaced by the result of the repeated measurement. The new measurement will then be the individual dry film thickness.

- If this individual dry film thickness does not meet the criteria it shall not be replaced.
- The number of replaced measurements shall be indicated in the test report.

Correction Values

If the Surface Profile is known and conforms to ISO 8503–1 correction values below can be used:

Surface Profile in accordance with ISO 8503-1:

- **Fine** – 10μm Correction Value
- **Medium** – 25μm Correction Value
- **Coarse** – 40μm Correction Value
- The specification/contract might require the determination of a specific correction value that is a correction value determined on the abrasive blast cleaned or roughened substrate with the particular dry film thickness instrument being used.

If the Surface Profile is not known and an uncoated sample is not available, a correction value of 25μm should be used.

- If a sample showing the surface profile is available and the profile is not in accordance with ISO 8503–1 the correction value needs to be determined.
- If a correction value is used it should be subtracted from the individual reading to give the individual dry film thickness in micrometres.

NOTE: The correction **value** is applied once on every reading no matter if the coating consists of a single layer or multiple layers

Expression of Results:

- The results of the measurements should be recorded as individual dry film thickness expressed in um or mm as appropriate.
- The mean dry film thickness belonging to an inspection area should also be given.

Acceptance/Rejection Criteria

For the acceptance of an Inspection area the following criteria must be fulfilled:

- The arithmetic mean of all the individual dry film thicknesses shall be equal to or greater than the nominal dry film thickness (NDFT).
- All individual dry film thicknesses shall be equal to or above 80% of the NDFT.
- Individual dry film thicknesses between 80% of the NDFT and the NDFT are acceptable provided that the number of these measurements is less than 20% of the total number of individual measurements taken.
- All individual dry film thicknesses shall be less than or equal to the specified maximum dry film thickness. If not specified see ISO 12944–5.

Note: For verifying Dry Film Thickness as given in the ISO 12944-5 acceptance criteria are as given therein.

NOTE: If the acceptance criteria above is not met, the Inspection area should be rejected.

- A Test Report is prepared at the end of each shift.
- Calibration should be performed at the beginning of each working shift, and then again when anything happens that could affect the accuracy of the instruments, such as dropping or leaving exposed to extreme heat or cold.

Tooke Gauge or PIG (Paint Inspector's Gauge)

ASTM D 4138: Standard Test Method for Measurement of Dry Film Thickness of Protective Coating System by Destructive Means: Test Method A - Groove Cutting Instruments

This method maybe used for both the measurement of the DFT and for examining defects in protective coating systems. The instrument has three cutting blades of different angles, which when used to cut through the coating system down to the substrate leaves a definitive angle on one side of the cut.

The instrument has a 50 X battery powered magnifier scope that must be placed over the cut in the coating.

Tooke Gauge

Depending on the cutting tip used the reticle contained inside the 50 X microscope can be used to count the lines in each layer of the coating system. The distance between each line in a cut made by a 1 X tip indicates a DFT of 25µm. The distance between each line in a cut made by a 2 X tip indicates a DFT of 12.5µm. The distance between each line in a cut made by a 10 X tip indicates a DFT of 2.5µm.

When used with heavy coatings such as non-skid epoxy coatings, a battery powered groove grinder such as a Dremel grinder may be used to cut a more precise angle in the coating system. ASTMD 4138 Method C involves the use of a specific angle tip drill (Saberg drill). Measurements are again made with a 50 X microscope. Each line on the microscope reticle has a DFT value of 20µm.

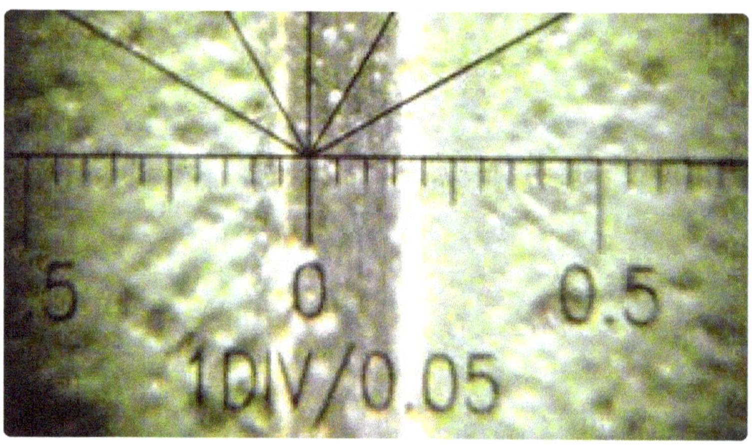

The view through a Tooke Gauge

Procedure

- Select a test panel or choose a site for the thickness measurement.
- Using an appropriate surface marker of contrasting colour mark a line on the surface approximately 2 inches long (51mm) where the thickness measurement will be made.
- Select a cutting tip based on estimated film thicknesses.
- If the thickness is unknown make a trial determination with the 2x tip.
- To cut a groove grasp the gauge with the studs and cutting tip firmly forming a tripod on the painted surface. Place the gauge at right angles to and about 2 inches (51mm) perpendicularly from a marked line.
- Draw the gauge across the paint film towards the body with the guide studs leading the cutting tip, and increase pressure on the cutting tip until it barely cuts into the substrate before it crosses the marked line.
- Take readings at the intersection of the marked line and the incision. Read by measuring on the reticle the distance from the substrate/coating demarcation up the longer machined slope of the incision to the upper cut edge of each respective coating layer of the coating system.
- Make sure that the smooth cut face of the groove is measured. The machined upper edge of the cutting tip usually leaves a less jagged cut. If multiple coats are observed individual thicknesses of each coat may be read. The actual coating thickness is derived by multiplying the reticle reading by the conversion factor for the respective cutting tip.

Adhesion Test Instruments

ASTM D 3359 "Standard Test Method for Measuring Adhesion by Tape Test" – Method A measures adhesion of coating films greater than 135μm. It consists of cutting an X of about 1.5 inches in length through the coating film then firmly placing a piece of special tape over the X and pulling it back 180°. The back side of the tape is observed to determine the amount of coating removed along the axis of the X cut.

ASTM D 3359 – Method a test:

Designation: D 3359 – 97

AMERICAN SOCIETY FOR TESTING AND MATERIALS
100 Barr Harbor Dr., West Conshohocken, PA 19428

Standard Test Methods for Measuring Adhesion by Tape Test[1]

1. Scope

1.1 These test methods cover procedures for assessing the adhesion of coating films to metallic substrates by applying and removing pressure-sensitive tape over cuts made in the film. Page No 1

3.1 *Test Method A*—An X-cut is made in the film to the substrate, pressure-sensitive tape is applied over the cut and then removed, and adhesion is assessed qualitatively on the 0 to 5 scale. Page No 1

Rating	Description
5A	No peeling or removal
4A	Trace peeling or removal along the incisions
3A	Jagged removal along the incisions up to 1/16" on either side
2A	Jagged removal along the incisions up to 1/8" on either side
1A	Removal of most of the coating from the area of the "X" under the tape
0A	Removal of coating beyond the area of the "X"

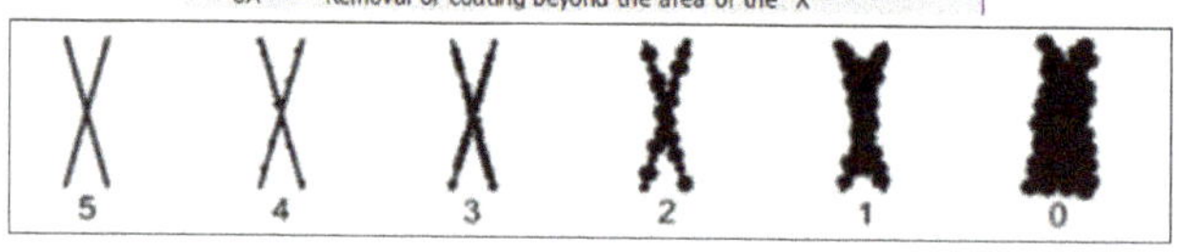

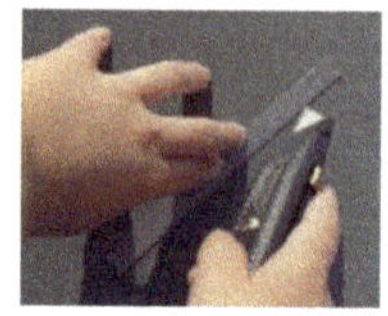

1. Cut "X" mark

2. Place 25mm wide Transparent Tape.

3. Remove the tape quickly.

Measuring Adhesion by the Tape Test Method

The rating system 0A indicates removal beyond the legs of the X, 1A indicates removal of most of the area of the X, 2A indicates jagged removal up to ⅛ in on either side of the X, 3A indicates removal up to 1⁄16 in along either side of the X, 4A indicates trace removal along the X, and 5A indicates no peeling or removal.

Method B measures adhesion of coating films less than 125µm DFT. For coating films less than 50µm DFT, 11 lines are cut with a sharp blade 1mm apart, in a cross-hatch pattern. For coating films between 50–125µm the cuts are made 2mm apart in a cross-hatch pattern. The same type of tape as Method A is then firmly placed over the cuts and pulled back 180°.

A 5B rating is where none of the squares in the cross-hatch pattern are removed. 4B indicates small flakes removed, but less than 5% of the area is affected. 3B indicates small flakes removed from 5% to 15% of the area. 2B indicates flakes removed from 15% to 25% of the area. 1B indicates flakes removed from 35% to 65% of the area. 0B indicates more flakes removed than 1B.

Procedure:

- The two cuts are each made in a length of about 40mm with sufficient forces to penetrate the coating film.

- The angle of intersection must be between 30° and 45°.
- Remove two laps of pressure sensitive tape and discard. Remove an additional length of tape about 75mm in length.
- Place the centre of the tape on the intersection of the cuts with the tape running in the same direction as the smaller angles.
- Smooth the tape into place by finger and then use an eraser and rub firmly.
- Within 90 +/-30 seconds of application, remove the tape by seizing the free end and pulling it off rapidly back upon itself at as close to an angle of 180° as possible.
- Inspect the X-cut area for removal of coating from the substrate or previous coating and rate the adhesion in accordance with the following scale below:

5A – No peeling or removal
4A – Trace peeling or removal along the incisions or at their intersection
3A – Jagged removal along incisions up to 1.6mm on either side
2A – Jagged removal along most incisions up to 3.2mm either side
1A – Removal from most of the area of the X under the tape
0A – Removal beyond the area of the X.

ASTM D4541-17: Standard Test Method for Pull- Off Strength of Coatings using Portable Adhesion Testers" – This is a widely used method that produces a measurement of the tensile forces required to remove a fixture or dolly that has been adhered to the coating with chemical adhesive, usually a two-component, 100% volume solids epoxy. Other quick curing adhesives may be used; however, the accuracy may not be as good as epoxy adhesives.

The dolly is glued onto the coating surface and the adhesive is allowed to cure. After the proper curing time, a mechanical device is used to pull on the dolly until more tensile strength is applied than either the adhesive strength of the glue or the coating system is exceeded. At that time, the dolly will pull off, and a reading of the pull-off strength in psi can be obtained.

This can be compared with the pull-off value that the coating manufacturer has stated to be appropriate for each particular coating, or it may be compared with a value that is required by the coating specification. In either case it will be reported, not only in its psi value, but also as to where the failure occurred. This can be glue failure where the tensile strength applied exceeds the adhesive strength of the glue that was placed on the dolly.

It can also be the psi at time of failure when one or more coats of paint separated from the substrate. There are also instances where the tensile strength of the mechanical device is applied up to a certain amount. If the dolly does not pull off at that point, the test is considered complete and a report is made indicating a passing test. Commonly used pull off adhesion instruments are Elcometer 106, Hate Gauge and Patti.

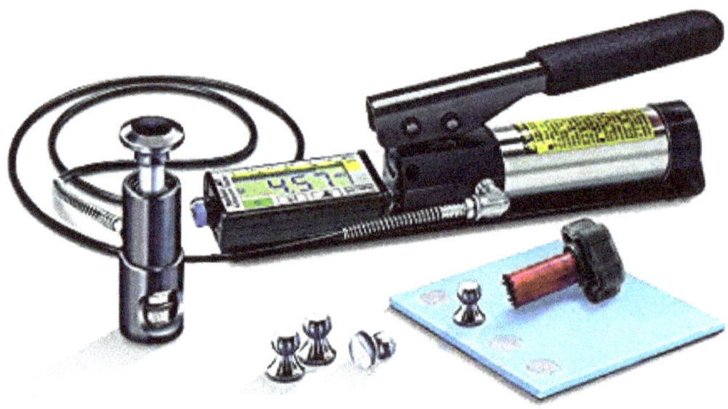

Pull Off Adhesion Tester

Pull off Test for Adhesion: ISO 4624

- Principle of conducting the test. The paint system under test is applied at uniform thickness to flat panels (the test is normally carried out on the object/structure that has been painted) of uniform surface texture. After drying/curing the coating system, the dollies are bonded directly to the surface of the coated cured panel using an adhesive.
- After curing of the adhesive, the bonded dolly assemblies are placed in a suitable tensile tester. The bonded assemblies are subjected to a controlled tensile test (pull off test) and the force required to break the coatings/substrate bond is measured.
- The test result is the tensile stress necessary to break the weakest interface (adhesive failure) or the weakest component (cohesive failure) of the test assembly. Mixed adhesive/cohesive failures may also occur.
- Tensile Tester: The tensile stress shall be applied in a direction perpendicular to the plane of the coated substrate and shall be increased at a substantially uniform rate, not greater than 1Mpa/s2, such that failure of the test assembly occurs within 90 seconds.

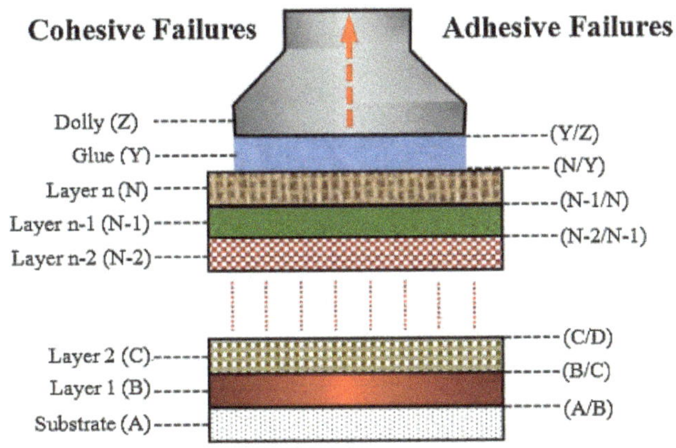

(a) The cohesion failure and adhesive failure according to ASTMD4541-09 standard.

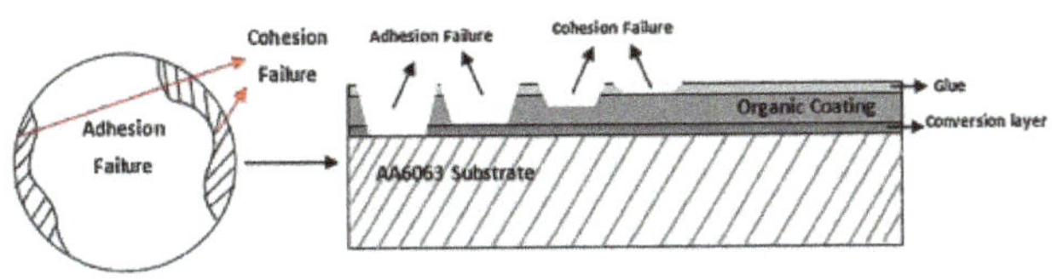

(b) The schematic drawing surface and cross-section of coating layers/metal substrate after pull-off adhesion test.

NOTE: Instead of a tensile tester, other types of pull off adhesion testers (mechanically, pneumatically, hydraulically or hand driven) may be used provided that they give similar results. The type of instrument shall be reported in the test report, because hand driven/ mechanically/hydraulic instruments are reported to produce widely different results.

NOTE: The results can be influenced by the test assembly used. Furthermore the results are not reproducible unless coaxial alignment of the tensile force is ensured.

Test Dollies: Each consisting of a steel or aluminium faced cylinder specifically designed to be used with the tensile tester. Each dolly has a rigid flat face for bonding the adhesive/ coating at one end and a facility for connecting the pull off tester at the other.

Each dolly has a nominal diameter of 20 mm and sufficient thickness to ensure freedom from distortion during the test. It is recommended that the length of each dolly is not less than half its diameter.

- The faces of each dolly should be machined perpendicular to its axis before use. Dollies with a diameter of 7 mm may be used when the method for testing adhesion from one side only is used. If dollies of 7 mm in diameter are used, 10 measurements should be made to improve the precision. The diameter of the dollies is to be reported in the test report.

- Centring device: for ensuring proper coaxial alignment of the test assembly during the adhesion process.
- Cutting device: such as a sharp knife for cutting through cured adhesive and the paint coating to the substrate, round the circumference of the dolly.

NOTE: Depending on the mechanical properties of the paint system cutting through the cured adhesive and the paint film to the substrate can have a big influence on the adhesion of the paint system. It is permitted if specified or agreed between interested parties, not to cut when the paint systems are less than 150µm in layer thickness. If cuts around the dolly have been made, this shall be mentioned in the test report and the type of cutting tool also mentioned.

- For convenience the below following scheme may be used to describe the results observed:

A-	Cohesive failure of the substrate
A/B	Adhesive failure between the substrate and the first coat
B	Cohesive failure of the first coat
B/C	Adhesive failure between the first coat and second coat
-/Y	Adhesive failure between the final coat and adhesive
Y	Cohesive failure of the adhesive
Y/Z	Adhesive failure between the adhesive and the test cylinder.

Holiday Detection:

It is critically important that coating systems intended for splash zones or immersion in corrosive liquids be pin hole or holiday free before being placed into service. It only takes one discontinuity in the applied film to allow the liquid (electrolyte) to reach the substrate. When that happens a corrosion cell is formed, the aggressiveness of which is determined by the liquid solution and the temperature of the environment.

There are a number of commercially available instruments to detect these discontinuities immediately after the coating has cured, thus allowing them to be repaired prior to going into service. There are both low-voltage and high-voltage holiday detectors available from a variety of sources. The low voltage detectors operate on 9-V batteries and are capable of detecting discontinuities in coating films of up to 250µm with only clean freshwater used as the electrolyte in the sponge.

For film thicknesses from 250µm to 500µm it is necessary to add a wetting agent to the water. For film thicknesses in excess of 500µm it is necessary to use either a battery operated DC high voltage holiday detector or an electrically operated AC high voltage holiday detector. In either case, it is necessary to adjust the voltage according to the DFT of the coating film being tested.

The most common setting is 100 V/mil but some specifications require 125 V/mil. These instruments should not be set beyond the recommended volts per Mil, or the spark created when a pinhole or a holiday is detected can burn or char the coating. These instruments can actually cause a pinhole or holiday when the voltage is set too high. Holiday detection should not be conducted after a coating film has been placed into service. The absorbed moisture will give erroneous readings, indicating pinholes or holidays where none exist.

Procedure for use following the NACE Standard RP0188-99 - Low Voltage

Sufficient drying or curing of the coating should be allowed prior to conducting a test. The length of time required for drying or curing can be obtained from the coating manufacturer. Solvents retained in the coating film could produce erroneous indications.

- The film thickness of the coating is to be measured with a non-destructive dry film thickness gauge. If the coating film exceeds 500µm then the procedure for the high voltage spark tester should be used.
- A low voltage wet sponge tester should not be used for determining the existence of discontinuities in coating films having a total thickness greater than 500µm due to the inaccuracy and lack of sensitivity of low voltage wet sponge testers.
- The ground wire from the tester ground output terminal must be attached to the conductive substrate and positive electrical contact should be ensured. The exploring sponge lead should be attached to the output terminal.
- The sponge should be saturated with tap water. The sensitivity of the test may be increased by adding a low sudsing wetting agent. The sponge should be wetted sufficiently to barely avoid dripping whilst it is moved over the coating.

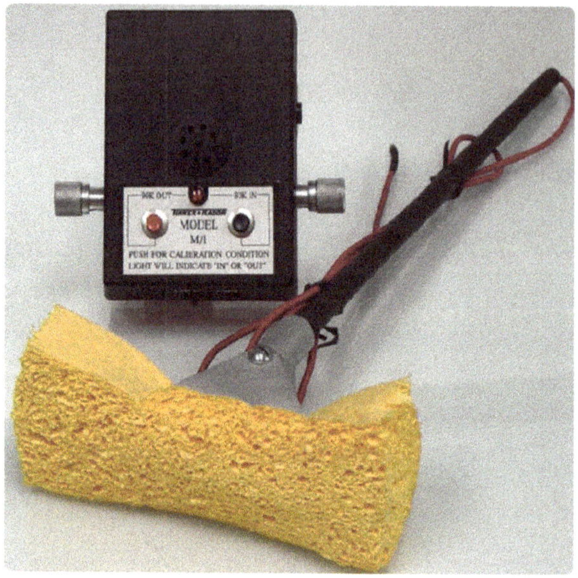

Low Voltage Wet Sponge Detector

- If a wetting agent is used, it must be completely removed by rinsing the holiday area prior to the repair. Additives can leave contaminants on the surface that can interfere with adhesion of topcoats or repair coats.
- If a test is conducted between coats of a multicoat system a wetting agent should not be used.
- Sodium chloride should not be added to the water because it can cause erroneous indications of discontinuities. The salt after drying on the coated surface can form a continuous path of conductivity. It also interferes with intercoat adhesion of additional coats.
- A bare spot on the conductive substrate should be contacted with the wetted sponge to verify that the tester is properly grounded. This procedure should be repeated periodically during the test.
- The sponge should be moved over the surface of the coating at a moderate rate of approximately 0.3m/s using a double pass over each area. Sufficient pressure should be applied to maintain a wet surface.
- Discontinuities that require repair should be identified with a marker pen that is compatible with the repair coating or one that is easily removable.
- To prevent telegraphing, care should be taken to ensure the solution is wiped dry from a previously detected discontinuity before continuing the test.

High Voltage Holiday Detector

High voltage holiday detectors are suitable for determining the existence of discontinuities in coating films of all thickness. The coating manufacturer should be consulted for proper test equipment and inspection voltage.

- **CAUTION:** Coatings that are applied at thicknesses of less than 500µm may be susceptible to damage if tested with a high voltage holiday detector.

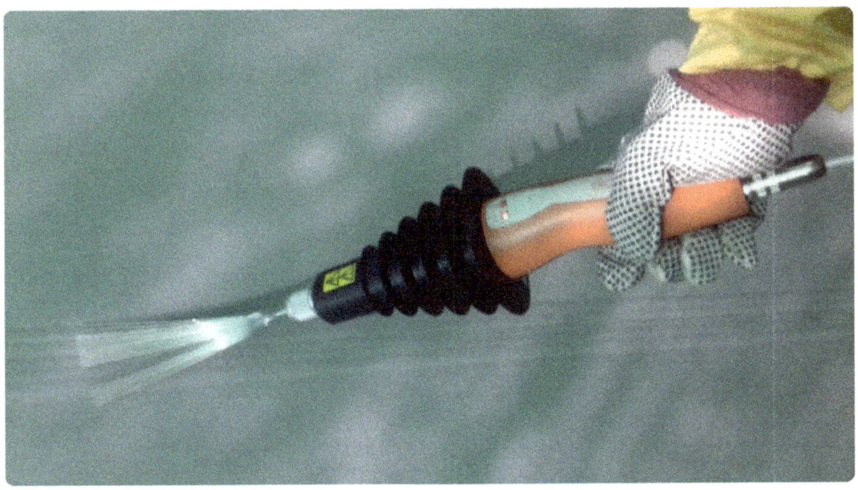

High Voltage Holiday Detector Probe

- A high voltage spark tester may be used to determine discontinuities in coatings on conductive concrete substrates. The conductivity of concrete varies depending on moisture content, type, density, and location of rebars. Conductivity should be tested by attaching a ground wire to the rebar or another metallic ground permanently installed in the concrete and touching the electrode to the bare concrete.
- If the metallic ground is not visible, the ground wire should be placed directly against the bare concrete surface and weighted with a damp cloth or wet sand filled paper bag. If the test indicates that the concrete is not conductive, determining discontinuities with a high voltage spark tester will be ineffective.

High Voltage Holiday Detection

Procedure for use following the NACE Standard RP0 188-99-High Voltage

Sufficient drying or curing of the coating should be allowed prior to conducting a holiday test. The length of the time required for drying or curing can be obtained from the coating manufacturer. Solvents retained in the coating film could produce erroneous results as well as an explosive environment.

- The thickness of the coating is to be measured with a non-destructive dry film thickness gauge. If the coating film is less than 500µm the procedure for low voltage wet sponge testing should be considered. Although the high voltage holiday detector is suitable for determining discontinuities in coatings films of less than 500µm, it is recommended that the coating manufacturer be consulted before using the test. Certain coatings can be damaged if tested with this equipment.
- The high voltage holiday detector should be adjusted to the proper voltage for the coating thickness to be tested. In selecting the inspection voltage sufficient voltage must be provided to break the air gap that exists at the holiday. The air gap varies depending on the total applied film thickness.
- Excessive voltage may produce a holiday in the coating film. The maximum voltage for the applied coating should be obtained by the coating manufacturer.
- The ground wire from the high voltage holiday detector ground output terminal should be attached to the conductive substrate and positive electrical contact should be ensured. When testing concrete surfaces the ground wire should be attached to the rebar. If the rebars are not visible the ground wire should be

placed directly against the bare concrete surface and weighted with a damp cloth or wet sand filled paper bag.

- Contact should be made with the exploring electrode on the conductive substrate to verify that the tester is properly grounded. This test should be conducted periodically during the testing of the coating.
- The exploring electrode should be moved over the surface of the dry coating at a rate of approximately 0.3m/s using a single pass. Moisture on the coating surface can cause erroneous indications. If moisture exists it should be removed or allowed to dry before the test is conducted.
- Discontinuities that require repair should be identified with a marker that is compatible with the repair coating or one that is easily removable.
- NACE Standard RP0188–99 should be followed whilst conducting this test.

High Voltage Holiday Detector with Brush Probe

CHAPTER TWELVE
Paint Film Defects

Amine Bloom

Name of defect: Amine Bloom. This defect has the appearance of a white film on the coated epoxy surface shortly after application. This milky discolouration often occurs with amine cured epoxy coatings.

What's the cause: It is caused by water on the coating film which has been applied in cold or damp conditions, especially at low temperatures, then a reaction takes place between the moisture and the curing agent. Applying the coating above 85% relative humidity and above will also have the same affect.

When the solvents evaporate from the coating and the temperature of the coating drops, this can lead to humidity condensating on the surface of the coating causing this flattened milky film. Amine curing agents are hygroscopic (absorb moisture), can react with moisture in the air to form ammonium carbamate.

How to repair: Amine bloom can be avoided if the coating is applied under more favourable weather conditions. The bloom is not detrimental to the coating itself; however, it will require removing prior to over-coating as it may interfere with adhesion. The bloom may be removed by a solvent rag if the area is only small. Scrubbing followed by high pressure washing is another method of removal; however, always check with the paint manufacturer's representative.

Alligatoring

Name of defect: Alligatoring. Resembles the skin of a crocodile or alligator and usually does not penetrate through the coating to the substrate as it is a surface reaction. This defect is also considered a checking/macro reaction and is caused by stress in the surface of the coating.

What's the cause – This is a stress related failure normally associated with a hard coating, possibly a top coat being applied over a softer mid/intermediate coat with the harder coating shrinking and pulling back exposing cracked areas. Alligatoring has also been observed on test panels where the coated surface has been force cured and with a rapid rise in temperature the curing process has been accelerated, which will cure the top coat exposing it to alligatoring and leave the under film soft. Natural drying oil-based paints such as alkyds tend to alligator as temperature fluctuates; also the application of a top coat before the intermediate coat/mid coat had dried can lead to the defect.

How to repair: Abrade with a fine sandpaper or dual action sander, clean, then recoat.

If the substrate is visible fully remove paint system.

Precaution: Adhere to paint manufacturer's recommendations. Avoid applying a hard coating over a softer intermediate/undercoat. Prevent from applying excessive film builds, especially coal tar epoxies. Ambient temperatures need to be carefully monitored before and during application.

Blistering

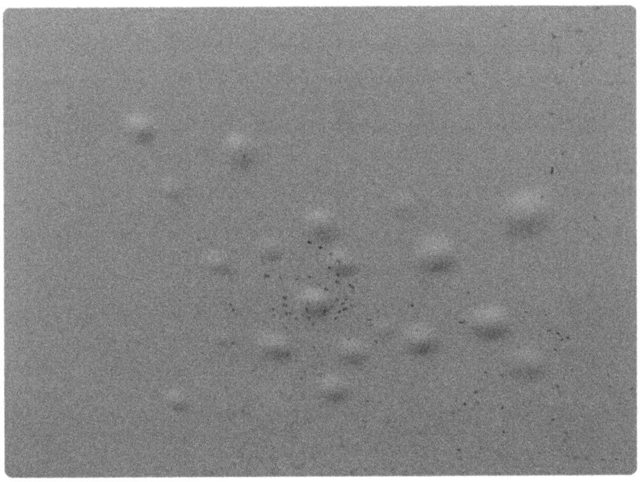

Name of defect: Blistering. The shape of the blister is often dome shaped/spherical and may be filled with a liquid substance or they may be dry.

Cause of defect: A number of coating issues may cause blistering; however, blisters will always occur where the adhesion is at its weakest. Two of the main reasons that blisters occur are due to solvent entrapment or osmotic blistering. Entrapped solvents are caused by excessively high film builds where the solvent being retained in the film is unable to evaporate and remain in the coating.

Over-coating too early with the next coat will also retain solvent leading to the formation of blisters. Osmosis is the spontaneous net movement of solvent molecules (water) through a semipermeable membrane (coating film) into a region of higher solute concentration (the salt contaminated substrate).

The process drives to equalize the solute concentrations on the two sides, but because salt cannot pass through the membrane (coating) it can never equalize. Water continues to permeate into the region. As the soluble substance dissolves under the paint layer, the pressure caused by the increase in volume can exert a greater force than the paint adhesion and cohesion forces, giving rise to the formation of a blister; the process called osmotic blistering. The blisters are first filled with water and later with corrosion products from the corrosion of the metallic substrate.

How to repair: In the case of entrapped solvent the entire surface requires to be removed and a new paint system applied.

Precaution: If blistering is due to soluble chlorides (salt) on the substrate, the entire coating system requires full removal followed by a freshwater wash to dissolve the water-soluble contaminant. A check for chlorides would also be advisable prior to painting. Avoid applying high film builds, adhere to strict over-coating intervals. Ensure surface for painting is chloride free.

Brush Marks

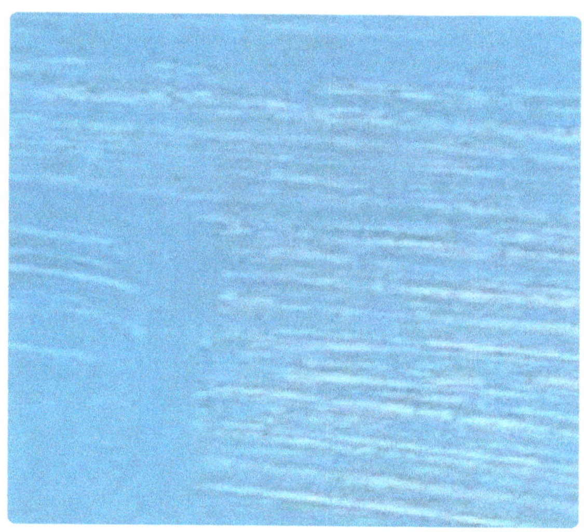

Name of defect: Brush marks are commonly known as ropiness and consist of valleys and furrows in the paint film where the coating has not "flowed or wetted". Poor brushing technique is the main cause of ropiness.

Cause of defect: High solid coatings and a coating with a heavy body which are difficult to apply by brush attribute to ropiness. Make sure the grade of the paint is "brush grade" and not "spray grade" as this will have an effect on the cosmetic appearance of the object or item being painted. Inadequate application technique and poor workmanship can also be the cause of brush marks in the coating.

A coating that has exceeded its "pot life" will be thick and viscous, so application by brush will be difficult with the material being "dragged" over the surface and unable to flow. Pinpoint rusting will occur in the furrows of the coating, leading to coating breakdown and failure of the system.

How to repair: Abrade the defected surface using a fine grade sand paper or a dual action sander, clean then recoat.

Precaution: Use the correct grade of coating "brush or spray grade"; if in doubt ask the paint manufacturer. Avoid over application and thick coats as two thin coats are better than one thick coat. Check technical data sheet for "induction intervals" and "pot life".

Bubbling

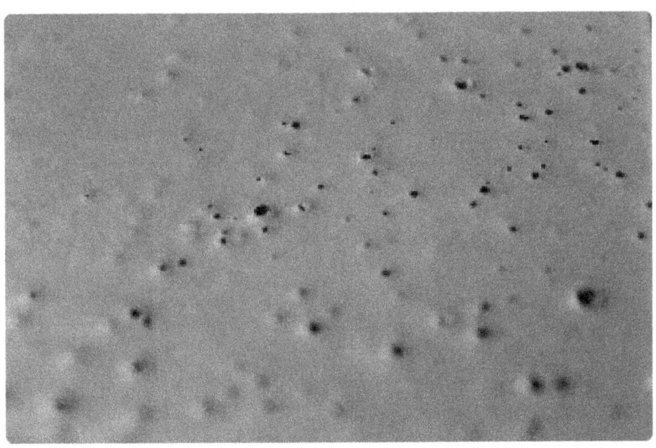

Name of defect: Air bubbles in a coating appear as small blisters; however, bubbling is more compact and bound together. Bubbling is associated with loss of adhesion. These are non-osmotic blisters.

Cause of defect: Coating application during high temperatures. The surface of the coating dries much faster than the body of the coating when it is exposed to direct sunlight. This prevents the solvents from evaporating. The coating then expands and creates a vapour pressure within the film, which in turn will create a bubble or multiple bubbles.

It is the vapour pressure which creates the bubble. Air entrapment due to over mixing is also a cause of bubbling as air is forced into the homogenous mix in the paint can. Applying coatings over a concrete surface will result in bubbling as concrete is porous and entraps air, humidity and moisture.

How to fix: If the bubbles appear during application then they can be laid off with a flat brush. If the coating is cured then the surface can be abraded with a fine sand paper to flatten and smooth the blisters then recoat, preferably in the shade.

Precaution: Avoid over mixing the paint material, use caution when applying coatings in high and low temperatures. Follow paint manufacturer's recommendations. When coating concrete tests can be conducted to measure moisture as humidity and vapour can be entrapped in the voids/porosity of the concrete surface.

Cratering

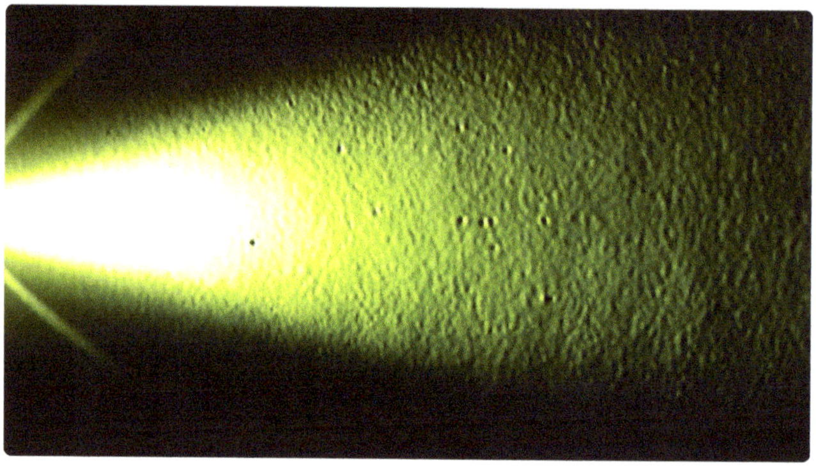

Name of defect: Cratering. Cratering is very similar to pin-holing; however, they are two entirely different coating defects. Cratering is a small depression in the film where the coating has not flowed or wetted the surface, displaying a volcano shaped defect normally not exposing the substrate. Coatings which have a high surface tension tend to suffer this type of defect.

What's the cause: Cratering is normally associated with an application malfunction or operator error (incorrect application technique). Contamination in the airstream can contribute to cratering with oil or water contaminating the surface. ASTM–D4285 Standard Test Method for Indicating Oil or Water in Compressed Air or the commonly known "White Rag Test" is a useful document in assessing such situations.

Solvent entrapment and air escaping are also accountable for cratering as the coating has not flowed sufficiently leading to depressions in the film. Application of high film builds directly on top of an inorganic zinc silicate primer causes extensive cratering.

How to repair: Abrade surface with a fine sand paper or dual action sander, clean, then recoat.

Precaution: Check for oil and water in the airstream to prevent contamination reaching the surface and use of the ASTM D4285 standard. Follow the manufacturer's recommendations as to spray tip size, use correct spray technique/application method. It may be required to abrade the surface or remove fully and apply a new coat. It is also advisory to apply a tie when over-coating a zinc silicate primer.

Cracking

Name of defect: Cracking. Cracking is a stress-related coating failure with stresses throughout the coating possibly down to the substrate; this is a serious coating failure. Cracking should not be confused with checking or alligatoring.

What's the cause: Cracking can be caused by a number of issues. Over application is a cause of cracking as high film builds stress the coating as curing continues. This can lead to a lack of adhesion and serious coating failure. As the coating ages and is exposed to the elements of wetting and drying, it is expanding and contracting together with a lack of flexibility.

Cracking is also familiar with alkyd and oil-based coatings which polymerize over time. All of these factors can contribute to cracking. The appearance of the cracking pattern may vary and it is often found in hard coatings applied over soft coatings. Zinc silicates have a high tendency to mud-crack if applied to thickly normally over 100μm. The recommended dry film thickness is around 75μm; however, this may change from product to product.

How to repair: In the case of zinc silicate primers, the entire coating needs to be removed and a thinner coat reapplied.

Precaution: As cracking can be caused by over application of coatings, wet film thickness measurements need to be monitored during spraying activities. Using the correct coating system will also reduce the chances of cracking with the use of proper resins, plasticizers and pigmentation.

The use of plasticizers will ease the coating, making it more flexible, and this will reduce the stress of the coating. Weathering and temperature change need to be monitored to avoid the formation of moisture on the surface of the coating. The Standard ASTM D 661-81 evaluation of the degree of cracking is used for the evaluation of this failure.

Chalking

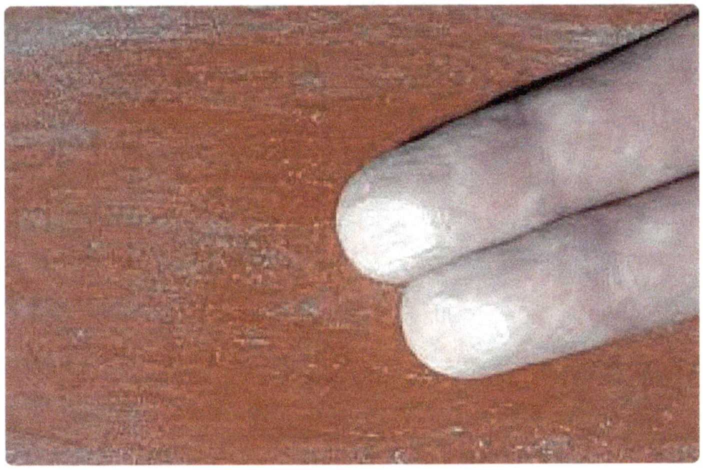

Name of defect: Chalking. Chalking is essentially a loose powdery layer on the surface of a coating. When the finger is rubbed over the surface a loose powdery deposit will be visible on the end of the finger displaying a fine powder.

What's the cause: Chalking occurs when the binder resin is disintegrated by the sun's UV rays, (actinic) humidity, dew and oxygen from the air. The binder resin near the coating surface disintegrates, leaving the pigment exposed on the surface.

How to repair: In most cases the surface can be cleaned by scrubbing and brushing to remove the powdery layer followed by washing with freshwater.

Precaution: If the surface is going to be over-coated it will require to be roughened to provide a suitable key for the next layer of paint. To avoid chalking is it important to use a paint pigmented with aluminium.

These provide a leafing effect like "fallen leaves on the ground" and are widely known for their performance of protecting the binder from UV or actinic rays of the sun that leads to the degradation of the binder, which in return leads to chalking. The plate like pigments improve adhesion and reinforce the binder resin with the leafing pigment being far superior than its counterparts. Polyurethane top coats are selected over amine epoxies due to this chalking effect. The tendency to chalking also varies with the type of coating used.

Checking

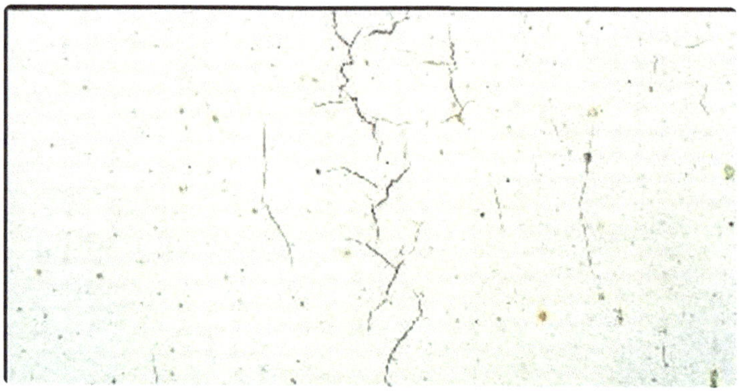

Name of defect: Checking. Checking is an age-related surface defect showing small breaks that do not penetrate through the coating to the substrate as it is a surface reaction.

Cause of defect: The older the coating gets it becomes harder and more brittle with the appearance of the checking pattern varying as it becomes older. Coatings applied in high film builds have a tendency to crack/check due to the stress within the coating. Checking also occurs when a hard coating is applied on top of a soft one.

How to repair: Checking is sometimes not found until years later as the coating becomes older and brittle. If the checking remains only in the top coat, it can be removed with a fine sand paper or dual action sander, cleaned, followed by a cosmetic coat.

Crazing

Name of defect: Crazing. Crazing appears to the naked eye as a loss of gloss, close examination under low magnification reveals a large number of minute cracks. These small cracks do not penetrate through the coating to the substrate.

Cause of defect: Excessive stresses in the paint film due to inadequate mixing of materials prior to application. Insufficient thinning or the wrong grade of thinners. Excessive film thickness, colour coat applied over inadequately dry or excessively thick under coats.

How to repair: Abrade the surface with a fine sand paper or a dual action sander, clean, then recoat.

Precaution: Always mix paint thoroughly, and use the correct amount of recommended thinner. Use correct spraying techniques. Apply paint in thin, web films and allow adequate drying time between coats. Only use approved additives.

Delamination

Name of defect: Delamination. Delamination is a loss of adhesion between coats in a multi-coat system and is common when repair or maintenance coatings are applied over cured or aged coatings.

What's the cause: Common within maintenance painting due to many factors such as incompatibility. The over-coating period could have been exceeded leading to an over cured and smooth surface. Polymerized coatings such as epoxies develop a hard, tough surface that must be abraded before over-coating to provide good adhesion between coats.

Contamination such as moisture or condensation, dust, dirt and grease can all lead to delamination. A chalked surface that has been over-coated will certainly lead to loss of adhesion due to the loose powdery layer still being present during painting. If the RH percentage has been too high during the curing of the epoxy coating, a phenomenon known as amine blush could occur. This amber sticky substance occurs if too much carbon dioxide is present on the surface together with a high RH percentage. This amber substance, if not removed prior to over-coating, will also lead to delamination.

How to repair: If applying over an aged or glossy coating make sure it is roughened so a key is provided and to ensure a good mechanical bond is achieved.

Precaution: Make sure the surface is clean and dry and free from surface contaminants, especially a heavily chalked surface prior to applying coatings. Adhere to manufacturers' paint data sheets.

Dry Spray

Name of defect: Dry Spray. A coarse and irregular surface consisting of dry particles of coating which have not flowed and sometimes are loosely adhered to the surface. Has a rough and sandpaper effect to the touch and often porous.

What's the cause: Holding the spray gun too far away from the object being painted. The solvent evaporates from the coating with the dry particles falling on the surface, resulting in a rough and uneven texture. If dry sprayed areas are over coated they will not "wet or flow" in and around the dry spray so failures such as pinpoint rusting will occur.

Poor application technique and poor gun adjustment will lead to areas of dry spray, therefore a continuous wet film is required. The problem may also occur when applying coatings in too hot climates, the solvent in the coating evaporates too quickly leaving a rough and coarse finish.

How to repair: Dry spray may be removed by abrading the surface with sand paper or a dual action sander.

Precaution: It is often difficult to make good the contaminated area without some type of grinding. Holding the spray gun perpendicular to the item being painted and using the correct gun distance will eliminate dry spray. Using the correct spraying technique with each pass over-coating the previous one by 50% will also eliminate dry spray on a surface.

Flaking

Name of defect: Flaking. Flaking is an adhesion related paint failure. This defect is associated with aged coatings that are hard and brittle. Once the coating cracks or breaks, it has a tendency to curl and lift from the surface, it may come off in large or small flakes.

Cause of defect: Aged coatings that are hard and brittle and continue to oxidize are prone to flaking. Another cause is that the over-coating interval has elapsed from application of one coat to another with no surface roughening taking place prior to over-coating.

Poor surface preparation on a zinc galvanized surface by not removing the zinc salts can also lead to flaking as in some cases remnants of contamination are often found on the back of the paint flakes. ASTM–D772–86 Evaluating the Degree of Flaking of Exterior Paint is a useful standard when conducting a coating conditioning survey or a maintenance plan.

How to repair: Ensure the correct coating system and generic type is used together with the correct levels of surface treatment. Surfaces with areas of visible flaking should be fully removed.

Fish eyes

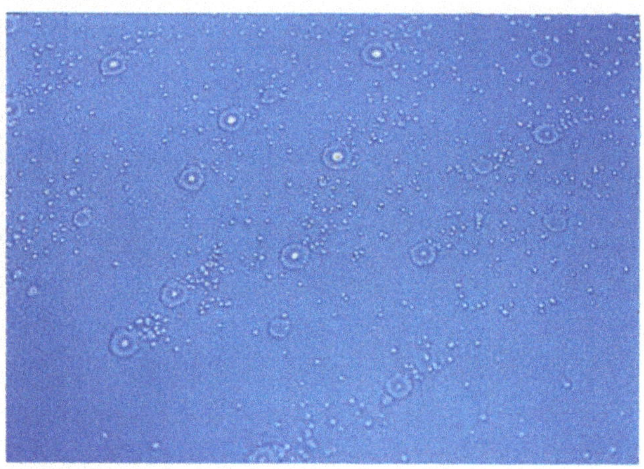

Name of defect: Fish eyes. Fish eyes are very similar to cratering; however, they are two entirely different coating defects. Fish eyes are small depressions in the film where the coating has not flowed or wetted the surface displaying a small spot in the centre of the failure.

Cause of defect: Fish eyes are normally associated with contamination in the airstream. Oil or water contaminating the surface from compressed air is the most common reason. Silicone contamination from grease or wax that remains on the substrate will also generate fish eyes.

ASTM–D4285 Standard Test Method for Indicating Oil or Water in Compressed Air or the commonly known "White Rag Test" is a useful document in assessing such situations. The most common reason that fish eyes occur is a faulty compressor, so the above standard for testing oil and water in compressed air should commence prior to any surface preparation and spray application. Any contaminants will show up as black spots.

How to repair: If the area is only small then abrade with a fine sand paper or a dual action sander so the fish eyes are ground flush.

Precaution: However, some of the dust may contain silicone particles which may be difficult to eliminate. If this is the case the coating will require full removal by blast cleaning.

Grit Inclusions

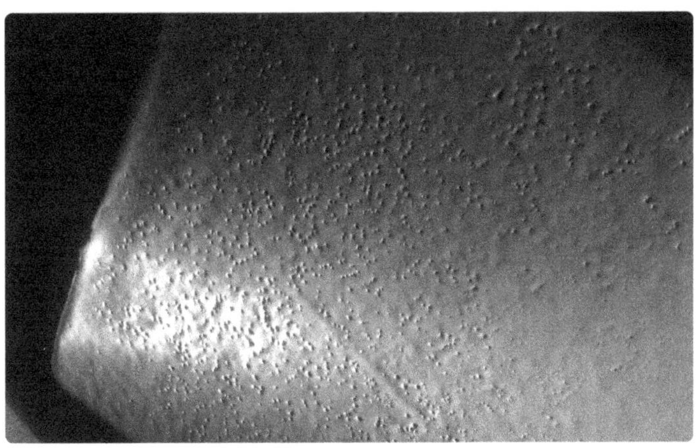

Name of defect: Grit Inclusion – Abrasive or blast medium sitting firmly in the dry film of a coating could be "airborne" or purposely painted over due to poor cleaning.

Cause of defect: Contamination of the paint film with blasting abrasive embedded into the coating. The surface feels rough to the touch like sandpaper. If the surface is not properly cleaned prior to coating application or airborne abrasives being blown onto the surface, this can lead to coating failure. If the blasting abrasive is metallic and is embedded in the surface it will create a corrosion cell and rusting will occur once in contact with an electrolyte. Make sure all surfaces are thoroughly cleaned and blown down prior to any painting activities.

How to repair: The contaminated area should be scraped to remove the embedded particles followed by a hand abrade with sand paper or a dual action sander to smooth over the coating.

Precaution: Painting operations should be separated from blast cleaning operations as cross contamination can be a major cause of airborne grit inclusion. Blast cleaning sheds should be fully encapsulated with no room for dust emissions to contaminate any coating works.

Gun Spit

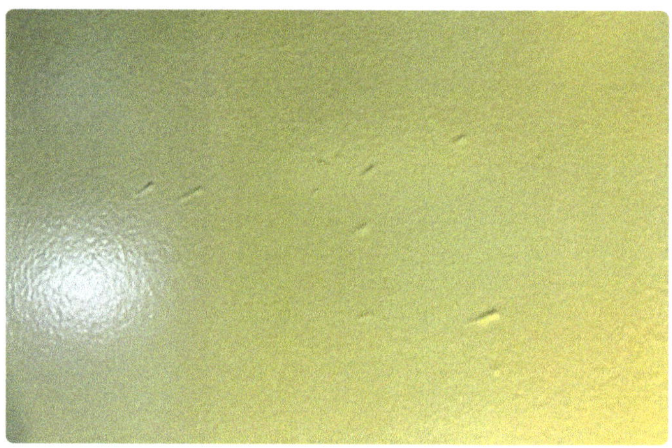

Name of defect: Gun Spit – Spits of paint on the surface of the item being coated.

Cause of defect: Gun spitting causes nodules or small lumps of coating on the surface of the item being painted as the "spray pattern or fan" cannot atomize properly due to blockages. These blockages of the spray tip will cause "spitting or chokers" due to a build-up or dirt, grit or dried paint.

With airless spray the tip can be reversed as this helps to clean out any blockages that may occur at the nozzle. Care must be taken when working with airless spray as high pressures can cause serious injury or even death due to paint being injected into the skin. Once injected into the skin medical help must be taken immediately.

Prevention: Gun spit can be removed by hand abrading the affected area with sandpaper followed by a touch up coat by brush, roller or spray.

Precaution: Maintain and keep the spray apparatus clean after each use. Use correct cleaning agents and gun-wash. Paint manufacturers will recommend tip/nozzle size for each product used which can be found on the paint data sheet.

Holidays

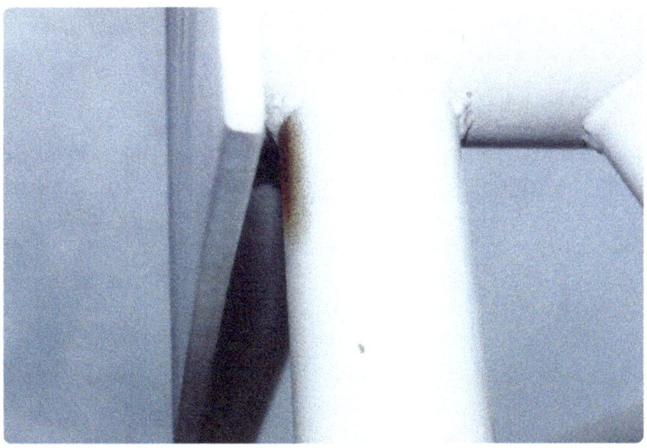

Name of defect: Holidays are essentially discontinuities or missed/thin areas in a coating system. These holidays can expose the substrate or the next layer of coating in that particular coating system.

Cause of defect: Mostly associated with poor application technique where the paint applicator has failed in applying the coating to the surface resulting in thin locations. These defects could be attributed to poor lighting or spraying in a tight area where movement is restricted due to supports or tank stiffening being obstructive. Holidays are generally associated with low dry film thickness so this type of failure can result in pinpoint rusting and undercutting.

How to repair: Improve application techniques and take more care.

Precaution: More adequate lighting or illumination maybe required when applying coatings and remember each pass should overlap the previous pass by 50% so a uniform and continuous film is formed. Improve quality control and supervision.

Lifting

Name of defect: Lifting. Lifting is the softening and expansion of a coating that has been recoated with another coating containing a stronger solvent.

Cause of defect: Lifting is normally associated with oxidative drying paints (oil-based paint). Once the paint is over-coated with another coating containing a stronger solvent, the first coat will begin to wrinkle and soften and after some time lose adhesion. The strong solvent will cause lifting and swelling with the solvent penetrating the coating attacking the binder molecules. Aged alkyds and epoxy esters are generally abraded with a dual action sander followed by a coat of epoxy mastic on maintenance painting programmes.

How to repair: Any coatings showing signs of lifting must be fully removed by blast cleaning.

Precaution: If the coating on the surface cannot be identified a simple test can be conducted. Simply wet a rag with a strong solvent (ketone) then leave the solvent soaked rag on the surface for 5 minutes. If after 5 minutes signs of wrinkling occur it is most likely an oxidatively cured coating.

Mechanical Damage

Name of defect: Mechanical damage is a condition where the coating is removed or damaged by dragging or scraping an item over the painted surface. As displayed in the above photo, the substrate is exposed due to the impact of the object hitting the coated item.

Cause of defect: Mechanical damage or transit damage as it is sometimes referred to is associated with the use of incorrect lifting or slinging techniques which will result in the damage of a coated item. If the straps or slings are not clean, they can contaminate the coated object leading to an unpleasing strap mark around the item being lifted.

Chains and unprotected forks on a forklift truck can result in catastrophic damage exposing the substrate and, in many cases, causing unnecessary and costly reworks. This can also lead to undercutting. Normally on large projects the coated items are preserved with polyethylene sheeting and heavy-duty nylon slings or straps are used to minimize handling damage.

How to repair: Adhering to the paint manufacturer's recommendations regarding handling times and curing for service intervals as this can minimize damages.

Precaution: The use of nylon slings or straps which are free from contaminants such as oil or grease and dirt can save time and effort.

Mud-cracking

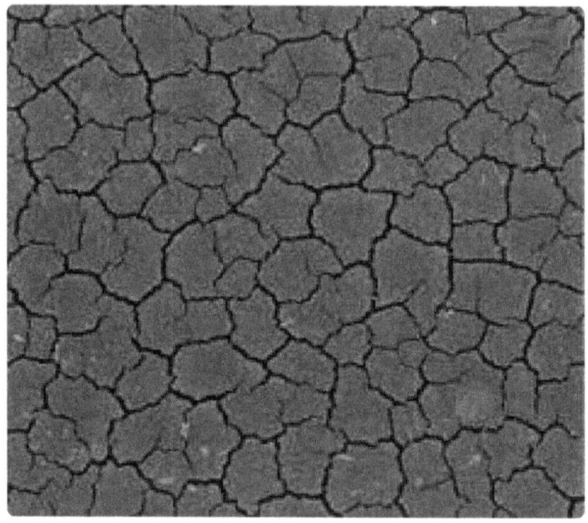

Name of defect: Mud-cracking. Mud-cracking has a similar appearance to alligatoring; however, this is a much more serious type of failure as the cracks penetrate all the way down to the substrate, therefore presenting a serious corrosion problem.

Cause of defect: Mud cracking occurs when inorganic zinc rich coatings are applied too heavily. Mud cracking occurs as soon as the solvent begins to dry out of the coating. The more rapidly the solvent evaporates the more likelihood that mud cracking will occur. The recommended dry film thickness for zinc ethyl silicate is around 75μm. If applied in higher film builds, more than 100μm, mud cracking will occur. The term mud cracking is used since the coating resembles a dried-out mass of mud.

How to repair: If mud cracking occurs with zinc ethyl silicates the system requires full removal by blast cleaning and a thinner coat applied.

Precaution: If fast drying conditions exist the applicator should avoid highly filled water-based silicates. Ambient temperatures must be strictly adhered to during application.

Orange Peel

Name of defect: Orange peel is an application associated defect which resembles the skin of an orange. This defect should not be confused with wrinkling.

Cause of defect: If a coating is applied by roller application and pressed too hard against the surface or rolled too fast in one direction then orange peeling will occur as the coating cannot flow. Using a roller with a heavy nap can also contribute to a stippled finish. The spraying application process can also result in an orange peeled texture with incorrect spraying technique/holding the spray gun too close to the surface causing insufficient atomization.

Orange peel may also be a result of too fast evaporation of the solvents in the coating. The coating is unable to wet and spread out on the surface, resulting in an orange peel effect. Excessive film builds also result in an orange peel finish. WFT comb measurements need to be monitored during application.

How to repair: Abrade the surface with sand paper or a dual action sanding machine.

Precaution: Be careful not to sand too far down and reduce the thickness of the coating system or even exposing the substrate. Since orange peeling is an application-related defect correct technique is essential.

Paying attention to the correct roller head selection and making sure the paint manufacturer's recommendations are strictly adhered to is also an important factor. The paint data sheet (PDS) will detail spray tip size and pressure at the nozzle. The tip size is measured in mils or thous and the pressure at the nozzle is measured in bar or psi.

Pinholes

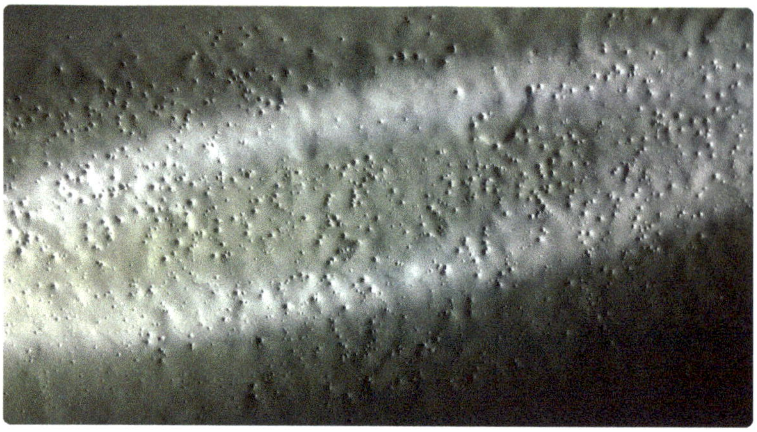

Name of defect: Pinholes. Pinholes are small holes or discontinuities in the film that can penetrate the coating all the way down to the substrate. Pinholes are weak spots in a coating system and need to be avoided. Care must be taken not to confuse pinholes with cratering.

Cause of defect: Pinholes. Listed below are a few reasons why pinholes occur. Pinholes are normally associated with thick or heavy coats where air is forced into the coating during application entrapping air bubbles. Pinholes often occur due to air and solvent escaping from the coating while the coating has not formed a coherent film.

When paint is mixed by mechanical methods (paddle mixer) over mixing the paint can lead to forcing air into material leading to pinholes developing during application. High ambient temperatures or spraying paint onto a hot substrate can lead to pinholing due to incomplete atomization.

How to repair: Applying another coat by airless spray will not eliminate pinholes as the paint will not be worked into the surface. Brush application is far better due to the superior wetting actions of the brush; however, this is unpractical on large areas.

Precaution: If the pinholes do not penetrate down to the substrate, abrade and recoat and improve the spraying technique keeping the gun perpendicular to the surface and at the correct distance from the object being coated. Pay attention to mixing techniques, not to over mix the material and avoid forcing air into it. Consult paint manufacturer for recommendations.

Pinpoint Rusting

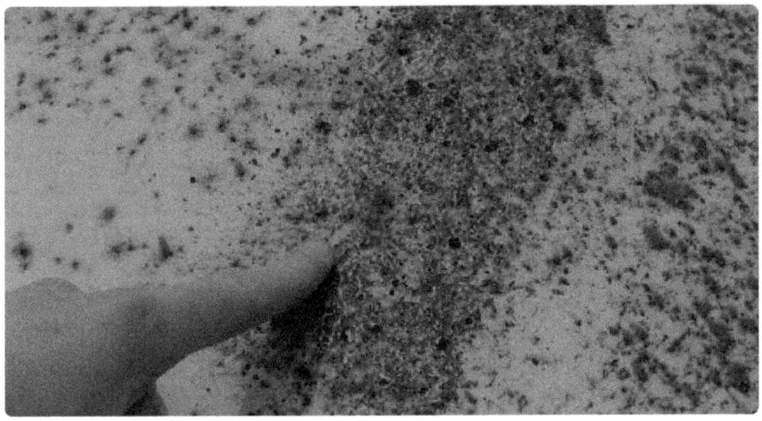

Name of defect: Pinpoint Rusting. Pinpoint rusting resembles spots of rust grinning through a painted surface or if large quantities are visible it will look more like a rust rash.

Cause of defect: Pinpoint rusting occurs when a primer coat is applied too thinly over a surface profile then left to weather over long periods of time exposed to the elements. It can also occur when the amplitude is higher than the thickness of the primer. (A high surface profile.)

If the amplitude is 85μm and the primer is only 50μm then pinpoint rusting will occur. Rogue peaks are also a cause of pinpoint rusting. These peaks stand out above the required amplitude and protrude through the primer coat. Rogue peaks are caused by blast cleaning in one area for too long.

How to repair: If the surface profile is visible through the primer coat then the coating requires full removal by blast cleaning.

Precaution: Surface profile needs to be measured with testex tape or needle gauge prior to the application of primer. DFT checks need to be conducted on the primer coat to avoid low dry film thickness.

Runs

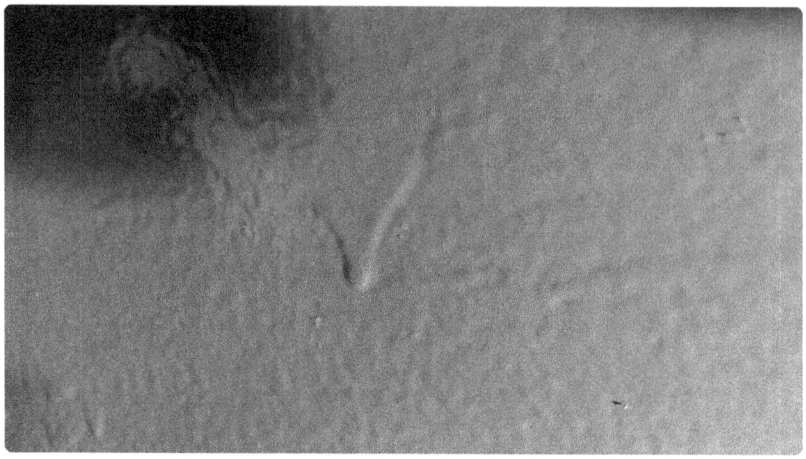

Name of defect: The running of paint often resembles that of a tear drop and is the result of applying a too heavy coat on a surface.

Cause of defect: If the airless spray gun is held too close to the surface or the applicator being in one spot for too long and not moving the gun perpendicular to the surface is a cause for running/sagging. Excessive amounts of thinners or the wrong thinner can lead to heavy sagging and running.

Small voids or pinholes are sometimes visible in the runs and sags as solvent and air are trapped; eventually these runs will crack and disbond, leaving the substrate open to failure. Applying a paint beyond its shelf life and improper mixing also leads to the running of paint. Runs and sags look unsightly and should be brushed out at the first opportunity.

How to repair: Abrade the runs with a fine sandpaper or dual action sander, clean then recoat.

Precaution: They can also be removed by blast cleaning at low pressures using a fine mineral abrasive followed by recoating. Take care and attention when applying coatings – try not to have too much material on the brush; if applying coatings by spray each pass should overlap the previous one by 50%, monitor WTF readings, adopt good spraying technique, make sure spray gun is not too close to the surface.

Soft Coatings

Name of defect: A pliable and soft coating often resembles the texture of "cheese" as it can be scraped away from the substrate quite easily.

Cause of defect: Soft coatings or "cheesiness" can be associated with the absence of the curing agent or the wrong curing agent added to the base. Insufficient mixing or the wrong mixing ratios can also attribute with paddle of the mixer not reaching the bottom of the pail to form a homogenous mixture.

The application of coatings on a cold substrate or cold ambient conditions can result in prolonged curing and poor film formation. Coatings can remain soft when they have been applied in a too high dry film thickness. If using a polyamide or amine curing agent, improper mixing can result in yellow blushing appearance and poor colour variations. Drying and curing times are also affected.

Repair of defect: If extensive cavities are found in the thick coating it should be assessed to remove the coating fully and a new coat applied. If no cavities are found in the film and curing continues, curing is still possible as long as the best curing conditions are provided.

Precaution: Ensure proper mixing of the material is executed and the correct curing agent is added. The correct mixing ratios are equally important as are the recommended thinners. Monitor climatic conditions during mixing of the material and application.

Undercutting

Name of defect: Undercutting. Undercutting is a measure of loss of adhesion. If there is a break in the coating, then corrosion will travel from underneath the coating away from the break, lifting the film and causing blisters and loss of adhesion. It is possible that the undercutting can travel many feet away from the break as shown in the picture above.

Cause of defect: Applying coatings over a corroded surface or due to lack of surface preparation with soluble contaminants present on the surface prior to painting. Mechanical damage is also a factor of undercutting and is seen in a variety of industries. Lack of maintenance/coatings especially on edges where the coating is thinner due to surface tension as the coating shrinks and pulls back leaving a thin film.

How to repair: Blast clean to Sa 2.5 followed by an appropriate coating system especially if used in an offshore environment. Freshwater wash surfaces prior to any form of surface preparation.

Precaution: Use a good coating system/maintenance plan including the use of an inorganic zinc silicate coating as these do not undercut due to the chemical/physical bond they have to the substrate. Epoxy coatings are normally chosen for over-coating inorganic zinc due to their excellent adhesion and undercutting values.

Water Stains

Name of defect: Water Stains. Pock marks and vacuoles caused by rainfall or moisture on the surface of the coating when freshly applied. This can also result in loss of gloss and a dulling on colour.

Cause of defect: Rainfall on a freshly applied surface causing crater-like marks on the surface before it has had sufficient time to cure or harden. A blooming effect can also occur if moisture or condensation is exposed during the curing process; this is normally associated with amine type coatings. Amine curing agents being hygroscopic (absorb moisture) can react with moisture in the air to form ammonium carbamate.

How to repair: If the area is only small and a cosmetic finish is not important then the area can be left as it is. If a cosmetic appearance is important then the area can be abraded with a fine sand paper, cleaned then recoated.

Precaution: Avoid rain or condensation during application. Make sure steel temperature is three degrees above the dew point temperature. Monitor environmental conditions throughout the day. If possible, make habitat/enclosed area and control the environment with dehumidification. Follow paint manufacturer's over-coating intervals and recommendations.

Wrinkling

Name of defect: Wrinkling. Wrinkling is the formation of furrows on the surface of a coating caused when the coating is applied too thickly.

Cause of defect: Wrinkling has more of a connection to oil-based paints like alkyds and epoxy esters as these coatings dry from the surface inward because oxygen is absorbed (oxidative coatings). If applied too thickly oxygen cannot penetrate the coating fully so the surface contracts and furrows and the coating beneath the wrinkling remains soft. Avoid applying alkyds in high and low temperatures and avoid thick coats.

How to repair: If wrinkling is evident in the coating, the coating will not dry completely. The soft coating requires removal by scraping followed by blast cleaning and the coating system reinstated.

www.ingramcontent.com/pod-product-compliance
Ingram Content Group UK Ltd.
Pitfield, Milton Keynes, MK11 3LW, UK
UKHW062003041026
15591UKWH00016B/1155

9 781914 195204